24,5⁰

T5-BCJ-262

Strategic Planning for Exploration Management

Strategic Planning for Exploration Management

Allen N. Quick
Neal A. Buck

International Human Resources Development Corporation • *Boston*

658.9622
Q6 s

To Donald M. Quick, Allen's father, who died June 28, 1982 — the day the initial paper, "Exploration Strategies as an Integral Part of Corporate Strategic Planning" was presented at the National AAPG Convention in Calgary, Canada.

Copyright © 1983 by International Human Resources Development Corporation. All rights reserved. No part of this book may be used or reproduced in any manner whatsoever without written permission of the publisher except in the case of brief quotations embodied in critical articles and reviews. For information address: IHRDC, Publishers, 137 Newbury Street, Boston, MA 02116.

Printed in the United States of America

Library of Congress Cataloging in Publication Data

Quick, Allen N. 1929–
  Strategic planning for exploration management.

  Bibliography: p.
  Includes index.
  1. Petroleum industry and trade — Management.
2. Petroleum industry and trade — Planning. 3. Prospecting — Economic aspects. I. Buck, Neal A., 1955–   . II. Title.
HD9560.5.Q48   1983   622'.18282'068   83-12710
ISBN 0-934634-66-1

Contents

UNIVERSITY LIBRARIES
CARNEGIE-MELLON UNIVERSITY
PITTSBURGH, PENNSYLVANIA 15213

Foreword

In all the industries in which I do management consulting, each manager considers his own industry to be truly "unique." Of course, each is different in some respects, and each has its own quirks and features. However, the similarities among industries far outweigh the differences. The critical dynamics and the management issues have a great deal in common. However, there are, I believe, two industries (or segments of industries) that have an important *critical* uniqueness that does distinguish them from the rest. One of these is the exploration for undiscovered natural resources, notably for oil and gas; the other is research. In these two industry segments, the competition is not nearly so much one firm against another as it is each firm against "nature," or—if you prefer—against the unknown.

This uniqueness not only sets these two industry segments apart from the rest, it also helps us to see what they have in common with each other:

— Pure scientific talent, ability, and genius have direct commercial value.

— We do not have the zero sum game of competition in the market place. A discovery by one firm does not usually perceptably lessen the opportunity of a "competitor" for a discovery. On the contrary, a discovery by one firm usually increases the knowledge of the whole industry, increasing competitors' opportunity for discovery.

— We see the source of continuing life for the rest of the firm. Oil and gas production, and the related downstream activities, can not long continue without a continuing and successful exploration effort. A firm relying on a continuing stream of new products or new technologies, such as an ethical drug firm, cannot long remain profitable without a stream of new discoveries continuing down the pipeline.

For management as well as for the strategist, this uniqueness has served as a challenge not often met. "How should you manage the exploration/research function?" is a popular topic for learned papers, seminars, and so forth; however, the answer to the question posed is not forthcoming. During the decades of the 1950s and 1960s, when — at least relatively speaking — times were good, money was cheap, and opportunities seemed to be unlimited, many firms avoided the issue by "throwing money at the problem." The popular style was to "hire good people, give them good tools, and let them do good things." Some firms still follow this approach today, both in exploration and in research. The clearest evidence is a firm's resource allocation to these activities through such devices as percent of sales, historical trends, or percent of cash flow generated from past successes. These methods are in stark contrast to an effort to determine the marginal benefit to be expected from a marginal expenditure in exploration/research, after taking risk and uncertainty into account, and after considering the opportunities implicitly foregone in other areas.

But, now I am beginning to sound like the strategist I am. We strategists have been grappling with a similar problem: How do you build a strategy for the exploration/reasearch function? A real breakthrough came about 25 years ago with the application of decision theory to these fields. Many firms today are pleased with this application and its results; others have tried it without success and abandoned its use. Still others avoid it on conceptual or cultural grounds. Even those who use it recognize a serious limitation: it increasingly relies on subjective judgments as the stakes get higher, i.e., as the exploration moves toward frontier areas or as the research moves toward basic areas. In exploration, by definition, one has no experience with frontier areas; in research, as one moves toward basic (i.e., "pure") research, one has little idea of what might be discovered. This is not an intrinsic limitation, since all strategy relies finally on subjective judgments. The issue in the application of decision analysis is whether to objectify and quantify at one stage the subjective judgments.

In any case, at best, decision analysis is a powerful tool in the formulation of strategy. *It is not a substitute for strategic planning.* Most of the other tools in the strategic planners' tool kit simply do not apply to these two industry segments, because of their unique quality. Competitive analysis, strengths and weaknesses, share of market, and rate of market growth are dull tools when we try to apply them to the formulation of exploration strategy. They become even more irrelevant when they are attempted against the even more critical strategic areas of control and implementation.

This is the situation, and Allen Quick and Neal Buck have picked up the challenge relative to exploration. I know of no other such serious attempt to take what we know about strategic planning and what we need in the management of exploration, and wrestle the resulting issues to the ground. I believe this work will become seminal in the application of strategy to the management of exploration. (Perhaps a similar effort will be forthcoming in the research field.)

The late Dr. John G. McLean probably did as much as anyone to bring modern management techniques, especially strategic planning, into the petroleum industry. As president and chief executive officer of Continental Oil Company, he said, "We can make money doing anything, but the way to score big is to find natural resources." (*New York Times,* June 28, 1970.) This is as true today as it was then, and it is much more difficult. Good science, good tools, good people, and money are just not going to be enough to yield commercially successful exploration in the long term. I am convinced, along with the authors, that successful exploration in the future will require good management and good strategy. Their statement that "In the future, only the exploration entity that follows good management practices will survive," may be only slightly too strong a statement. I would say, rather, that the firm with a good strategy, well managed and seriously controlled and implemented, will have a competitive advantage over those that do not.

The authors have properly scoped their work for and targeted it at the exploration sector of the petroleum industry. This was necessary to give their work cohesion and integrity. They probably did not expect what I perceive to be the by-products of such an emphasis. Their work turns out to be a useful reference work for those outside the industry, but interested in its workings. Perhaps it is so useful as such a reference for the very reason that it was not intended as one. For example, their description of industry data and data bases and of data analysis, interpretation, and use are helpful not only to exploration managers and strategists, but also to bankers, investors, etc. The description of industry structure, dynamics, environment, and traditions would be of interest to anyone wanting to learn more about the industry.

The coverage of risk management is, in my view, necessarily quantitative; however, enough description is provided so that the nonmathematician need not be frightened away. In fact, the quantitative approaches presented might well be applicable to fields other than exploration (e.g., research?). Assuming that we are making the kind of progress I perceive in dealing with the *implementation* of strategy, the next most needed task in enhancing the value of the discipline of strategic planning is to develop ways of considering risk and uncertainty in a serious way. We strategists say that increasing uncertainty is an underlying cause for the need of strategic planning, and then we mostly wave our arms when we talk about "risk." We have to learn to talk about risk with our hands in our pockets! That is, we must create the kind of disciplined, structured, replicable methodologies that we did relative to cash flow, strategic business units, competitive analysis, etc. We have a start with such useful tools as decision analysis, scenario planning, etc. Perhaps the industry of exploration (and research?) will prove to be the vehicle for the next breakthrough, whereby we could deal explicitly and evenly with risk. Perhaps it would be transferable to other industries, where risk and uncertainty are great, but are less well understood and appreciated; therefore, they are much more dangerous than they are in exploration. I personally believe the exploration industry just might very well be that vehicle—and very soon.

Ben C. Ball, Jr.
President, Ball & Associates
P.O. Box 158
Cambridge, MA 02142
June 29, 1983

Preface

This book evolved from a paper given at the National AAPG Convention in June 1982 and later published (*Oil and Gas Journal*, Sept. 27, 1983). When the call for papers occurred early in the fall of 1981, there were only slight signs of what was in store for the oil industry. These signs, however, were very definite and definable to someone who had lived through the past thirty years in the industry. The purpose of the initial paper was to give explorationists, particularly exploration management, an overview of strategic planning and, specifically, to show how exploration strategies could be included in their company's strategic planning, which would help them survive in their business in the uncertain future.

Most exploration managers have been exposed in management short courses to the theories of planning, and many are regularly exposed to the nuts and bolts of planning in their operational reporting to corporate management. However, many still have no way of relating their operations and informal planning to formalized corporate planning systems. This book was written to help bridge the gap between setting exploration strategies at the operational level and corporate strategic planning.

Allen is a geologist by training and has been involved in economics and planning for the majority of his career. For the last ten years, he has worked as an operational planner in an exploration office working directly with exploration management. Neal, an industrial engineer by training, has several years of computer-related experience with planning and petroleum industry applications. Although texts on strategic planning are generally written by management theorists from industry or by academics, this book is basically a view from the "bottom" or operational viewpoint.

We hope to make strategic planning understandable to exploration management and in turn relate exploration as a business to financial and strategic planning and corporate management. Methods and techniques are provided that can be used to integrate exploration strategies and corporate strategic planning. Only in this way can exploration operate as a business, which is necessary for long-term survival. Planning is a managerial responsibility that has long been neglected at all levels of management. Perhaps this book will provide a bridge, so that objectives, goals, and strategies of the corporation can be followed at the operational level. Also, the feedback resulting from the process will provide a basis for realistic planning at the corporate level.

In Bob Tippee's article entitled, "Industry Adjusts Strategy, Trims Operations to Handle Recession" (*Oil and Gas Journal*, Nov. 8, 1982), Texaco CEO, John K. McKinley described successful energy companies in the future:

1. *They will be flexible, able to develop effective strategic business plans while retaining the ability to implement new ideas quickly and adapt to new political, economic, technological, and business conditions.*

2. *They will be staffed by technically oriented people who are highly selective in their activities and people who are accountable for the projects they recommend. No longer will management throw money at a generally promising idea. Time and funds are simply not available, nor are they ever likely to be again.*

3. *Successful energy companies will have fewer but better qualified people, people who understand the most effective relationship between business and technology.*

McKinley's words set the future tone of the oil industry. Closer coordination of planning at all levels will be required to include exploration business in corporate strategic business planning. All planning will have to be flexible enough to react to changes in the environment. In the future, capital allocation processes and control systems will receive much attention. Technical people will need to adapt a business perspective to survive in the uncertain future facing exploration. We hope this book will help explorationists and planners alike adapt to the industry transition.

Denver, Colorado
May 1983

Allen N. Quick
Neal A. Buck

Acknowledgments

Special thanks to Hugh Watson who secured Cities Service Oil & Gas Corporation management approval to publish the book. We especially would like to thank Cities Service for providing a work atmosphere that has, over the many years, allowed employees to be creative and develop tools and ideas as embodied here. Additionally, we appreciate the use of material from work done under company auspices. Credit is due to David A. Jones and Jack H. Kelsey for their contributions to strategic analysis.

Other individuals within Cities Service that deserve special thanks are Art Grove and his staff for some of the illustrative material, as well as Art's help and advice on other matters, to Linda Hill and Paige Graening for their review of the bibliographic references, and to Candie Doolittle for her efforts and patience in word processing the manuscript.

Individuals outside of Cities Service that deserve special recognition are Ben C. Ball for instilling valuable ideas and for his help and encouragement during the preparation of the manuscript and Richard C. TenEyke for his early guidance in providing an understanding of the corporate side of strategic planning.

Special thanks to Jerlene Bright and Al Swartzkoph of the University of Oklahoma for preparing Appendix A describing the Petroleum Data System. Thanks also to Mike Thacker, Tom Dougherty, and Philip Stark of Petroleum Information Corporation for their cooperation in furnishing Appendix B describing the use and methodology for P.I.'s data bases. Thanks also to Donald Taylor, Linda Bacigalupi, and James Kent of FUND for Appendix C concerning Social Risk Factors and FUND's pending data base of these factors.

Special acknowledgment to Rocky Burnett for drafting many of the text illustrations.

Much gratitude (and credit) to Lora Buck for editing our original manuscript, certainly no small task, and to Dr. John A. Pederson for his consistent review of the manuscript.

Last, but not least, we wish to thank our wives, Mary and Lora, for putting up with us over the time that it took to prepare the manuscript and to console them for the lonely nights and weekends and undone chores during this period.

1

Introduction

Changes in Economic Environment of the Oil Industry

The economic conditions facing the oil industry changed drastically in 1982. The euphoria that surged through the energy industry in 1981 was quickly dampened in early 1982. Those people who thought the boom would go on forever were suddenly disappointed. The erosion of the world's crude prices due to the surplus of oil production, which caused reduced cash flows, was the immediate indicator of a sharp change in the economic environment. Companies of all sizes felt this change almost immediately. The next development was the softening of gas prices and the surplus of gas in certain markets. Currently, in the gas areas, there are problems in getting new wells hooked up, with additional concern about pipeline takes from wells that are connected. However, the real disasters occurred in the "high priced gas" areas such as the deep Anadarko gas play, where extremely high well costs required nearly ten dollars per thousand cubic feet to pay out the well economically.

The primary reason that most companies were caught short by the economic changes was the failure of forecasting. International affairs since the Arab oil embargo of 1973–74 have caused many unpredictable events. Most of the supply models used by industry and the federal government failed to predict that world petroleum demand in 1981 would be as low as 47 million barrels of oil per day. The forecasts in the early 1970s had set estimated demand for 1981 as high as 75 million barrels of oil per day.

The resulting drop in cash flow caused repercussions throughout all segments of the oil industry. Many of the small oil and gas companies went out of business. For the others, reduced cash flow brought about revision in strategic thinking and forced most companies to drastically overhaul current operations. As expected, the "knee jerk" reaction has been to shorten the planning horizons. The emphasis has quickly shifted from net asset values of reserves to the immediate cash flow. The industry, by being ahead of its own cash generating capabilities, was set up for a

1

tremendous cash drain. It had mortgaged its future against crude oil and gas reserves that were expected to steadily, even rapidly, increase in value.

Reaction to the sudden reduction in crude prices was varied. Hardest hit were the independent oil companies. They had borrowed money on the basis of constantly increasing product prices and in some cases reserve estimates that were not completely substantiated. The lending institutions suddenly realized that these loans were not fully secured. Some lending institutions failed under these conditions, and these failures had a repercussion throughout the entire banking industry. Some banking institutions reacted by demanding partial repayment of loans that were not completely secured. Prepayments equal to the difference between the banks' current appraisal of reserve values and the outstanding loan values were required. These prepayments put binds on small companies who had already experienced cash flow reductions due to lower product prices. In many cases, this condition resulted in bankruptcy.

Those oil companies that survived the initial onslaught were faced with the next problem—how to raise funds for future exploration. The equity market had all but dissolved for those firms that traded over the counter. The larger firms that had used various types of drilling funds found that the investment community had drastically changed. The "Scarsdale dentist" and other wealthy individuals were not looking for intangible write-offs for tax purposes because the tax laws had changed. They were now faced with investing "50-cent" dollars instead of "30-cent" dollars. The large public drilling funds that were so popular during 1979 and 1980 were also disappearing. The firms that depended on these types of financing were forced to reduce their drilling programs.

The major oil companies reacted to the price cut in various ways—cutting costs in all areas and deferring exploratory and development drilling. Many companies resorted to selling assets, including producing properties in some cases. The traditional funding for major companies is different from that of the smaller companies. Their sources can be split into four general categories: new equity issues, long-term debt, medium-term debt, and short-term paper. This type of funding will probably continue. However, there are concerns by some firms that large debt equity ratios may prevent them from increasing balance sheet debt.

Events like those in 1982 are not unique in the history of oil exploration. Drilling has always been dependent on oil and gas prices. The dips in drilling in the late 1950s and early 1960s were precipitated by lower prices and crude surplus. The Federal Power Commission-Phillips decision in the 1950s artificially depressed gas prices which in turn reduced efforts in gas exploration. The extraordinary aspect of the current downward turn is that it follows the two best years (1980 and 1981) that the oil business has experienced in its entire history. The expansion of the exploration segment of the business was unparalleled in the history of exploration, making it all the more difficult to adjust to the current turn in circumstances.

Figure 1.1 shows the oil price related to cycles of drilling since 1918 for all wells and since 1940 for exploration wells. To illustrate the interrelationship of crude price variability and drilling activity, the year-to-year change in crude price is expressed in percent. This method overcame the large swing in actual crude prices. Its value in percent was then plotted as either a positive or negative value so that the fluctuation of drilling could be related to crude prices. Interesting relationships and some history of the industry are illustrated. The instability of the crude market before the formation of the Interstate Compact Commission is evident in the 1920s and early 1930s. The effect of World War II is also visible. The postwar drilling boom following higher, stabilized prices is evident. Finally, the recent drilling boom is vey well illustrated.

There is a cultural element to the adaptation needed to adjust to this downward turn. Many of the individuals currently involved in exploration management have never before experienced difficult times. Those people who lived through the previous cycles and watched the expansion of the late 1970s with a practiced eye knew that the boom would not last forever; but not until the world economic conditions and the related glut of production occurred did the industry realize it had returned to the "good old days."

Exploration Viewed as Business
Rather than as a Function

Surely some good will come out of the current economic turmoil. The companies that survive will be financially healthier in the long run. The change in the marketplace has brought about periods of consolidation in which the strong bought out the weak. Many of the firms with little financial background that entered the exploration business with little or no planning experience will have perished. Institutions that have contributed increasing amounts of money to the industry over the past few years will have gained sophistication. This better informed community will force the oil industry to act with expertise in the areas of business and finance. This does not mean that the oil industry will turn into an actuarial exercise but that it will be looking for good management and good track records. In 1981 and early 1982 the ease of obtaining funds allowed many wells to be drilled that should not have been drilled.

It is obvious under present circumstances that exploration management will have to conduct its activity in a more businesslike manner. As Ben Ball [1], a strategic planning consultant and former vice-president of a major oil company, said "in the future it would be as inadequate to describe exploration as looking for oil as it would be to describe McDonald's as selling hamburgers."

In the past three years many bad business decisions in exploration have been

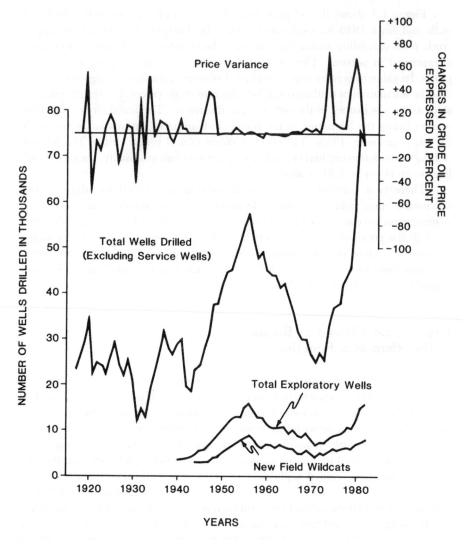

Figure 1.1. Drilling activity 1918–1982 for total wells and exploration wells versus crude price changes by years.

SOURCES OF DATA: well activity: world oil, crude prices: U.S. Department of Energy and Independent Petroleum Association of America.

rescued by the rapid increase in oil and gas prices. In the future, explorationists will not have the luxury of making poor decisions and still surviving. The traditional exploration strategies, as viewed by the explorationist, will need to be modified to include the company's business strategies, objectives, and goals. This modification is needed to make exploration a true business.

Strategic planning has become increasingly popular in the last fifteen years and has been successful in the product manufacturing business. In this type of business strategic planning deals with share of market, production cost, and growth areas. Strategic planning provides a method to help guide chief executive officers (CEOs) in deciding where they want their companies to be in the future. The strategic planning process itself is dynamic. In some segments of the petroleum industry, strategic planning is similar to that in the manufacturing industries because both have products, processes, and markets. Currently, there is a gap between strategic planning and exploration planning phases of the business. Long-term financial goals and predictability are not as definable in oil exploration as in other sectors of the petrolem industry, due in part to the complexity and inherent risks of the exploration process. In addition, more time is required to measure the results of an exploration program than is needed for other endeavors.

In the chapters that follow, the business aspect of strategic planning and its analysis and techniques are related to the exploration business. In the future, only the exploration entity that follows good management practices will survive. Current business conditions guarantee only the survival of the fittest.

Reference

1. Ball, B.C., Jr.: Letter to Author (Nov. 15, 1981).

2

Strategic Planning
Overview

What is Strategic Planning?

This overview attempts to give the explorationist the background necessary to understand the general strategic planning process. This chapter is not an attempt to exhaust all aspects of strategic planning. Many textbooks and articles have been written on strategic planning. The reading list at the end of this chapter suggests some additional material that may be helpful to those who wish to pursue the subject.

Planning in General

An understanding of the basic concepts of planning is helpful before entering into a discussion of strategic planning, so that these concepts can be included in the discussion. In general, planning is a continuous process of deciding what is desired and what can be done to make it happen. Planning requires a continuous attempt to adapt the company to a constantly changing environment.

Planning, often confused with other processes, is not budgeting or forecasting. Planning should not be used to predict the probabilities of possible events but rather it should be used to innovatively try to modify these probabilities. In turn, planning is not intended to eliminate risk but rather to select the risk to be taken.

"Planners" do not plan: managers plan. People who are called "planners" provide support for those who have decision-making authority (managers). Therefore, the chief planner for a company is its chief executive officer (CEO), who sets the pace for the planning effort. The CEO and delegated managers are the actual planners—they control the strategic planning process and the company.

Having defined planning, the actual reasons for planning should be noted. The purpose of planning is to aid present decision-making processes while keeping in mind possible future events. If management consisted of intuitive geniuses, there would be no need for planning. Because very few top managers have this gift, a formal planning system is required. The system provides management with data and alternatives of action that are a partial substitute for intuition.

Ewing [1] succinctly summarized the purpose of planning as follows:

1. *developing the objectives and goals for a company*
2. *projecting economic conditions that will affect its future*
3. *formulating alternative courses of action to reach the goals*
4. *analyzing the consequences of these alternatives*
5. *deciding which programs are most feasible in the light of limited corporate resources*
6. *devising methods for measuring progress toward a planned goal when a program has been chosen*

Before planning of any type is instituted, upper-level management must perceive a need for it. In most cases, organizational trauma seems to appear before planning is undertaken. If a company is flourishing it may seem that there is no immediate need for planning. Prerequisite to the introduction of successful planning is willingness, at all levels of management, to accept change. Successful planning requires innovation and creativity. Problems exist in developing and fostering an innovative organization that is receptive to planning. The majority of the staff in large established organizations are concerned with serving and maintaining the ongoing operation. This dominant group views the innovators as mavericks or as a deviant culture, and it will make efforts to stifle or hinder the innovators. Top management support is absolutely necessary to combat these tendencies and make planning successful.

There are inherent risks to management involved in undertaking formal planning. First, established management's past decisions will be analyzed and scrutinized and some poorly informed decisions will likely be exposed. Management must be ready for this analysis. If management feels secure in its decision-making position then scrutinizing company performance will not pose a threat. Also, if management is merely going through the motions of planning and not using the planning results for its decision making, this will soon become obvious. Managers who prefer to "shoot from the hip and then draw the target around the bullet hole" will certainly not be comfortable with a formal planning system.

Strategic Planning

Webster's *New International Dictionary* [2] defines strategy as "the science and art of employing armed strength of a belligerent to secure the object of a war." A more restricted definition is "the science and art of a military command, exercised to meet the enemy under advantageous conditions." Translating strategy into the terms of the business community, McNichols's [3] definition is: "The science and art of employing the skills and resources of an enterprise to attain its basic objectives under the most advantageous conditions." And according to Anthony, Dearden, and Vancil [4], strategic planning is "the process of deciding on the objectives of the organization, on changes in these objectives, on the resources used to attain these objectives, and on the policies that are to govern the acquisition, use, and disposition of these resources."

Strategic planning systems were introduced in the mid 1960s and were an outgrowth of long-range planning systems. These long-range planning systems were based on extrapolation of the past and lacked the flexibility needed in strategic planning. Strategic planning primarily deals with the contrivance of organizational efforts to direct the development of organizational purpose, direction, and future generations of service or products. It requires the design of implementation policies and strategies by which objectives of the corporation can be accomplished.

The base of strategic planning consists of three economic objectives according to McNichols: *profitability, growth,* and *survival.* These objectives are paramount in any business organization and will shape or determine its business objectives. Constraints against profitability and growth, coupled with the need to survive, are motivating factors that provide thrust for companies to expand or diversify. In the long run, these three economic objectives must become the overriding basic objectives of the managers of business enterprises in a competitive industrial society.

Structure of Strategic Planning and Policy Formation

The policy formulating process can be broken into four phases, which will define the structure of strategic planning: *formulation; implementation; organization and control;* and *reformulation — feedback and review* (see fig. 2.1).

Formulation Phase

In the initial stage of development of a business enterprise, the policy maker (the CEO) must appraise the skills and resources of the firm and determine the basic objectives for the organization. These objectives will shape the image and character of the company in the future and provide the mission and purpose for the firm. The mission statement for a firm is the keystone of the strategic planning process. In order to develop the mission and mission statement, some critical questions must be answered. What has been the organizational purpose? What exactly is the present business? Is this what the business could be in the future? What actions are needed to assure that the organization will realize its objectives for the future?

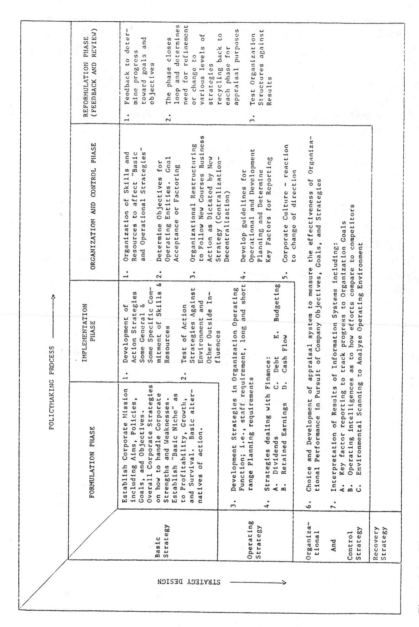

Figure 2.1.   The structure of strategic planning and policy formation.

SOURCE: Modified from T.J. McNichols [3].

These questions will identify the subject matter for the mission statement. The broadest choice in strategic planning must be made by the company when it defines its mission statement. It may be as simple as "we make widgets." The statement may be a very narrow description, which would be suitable for some firms. However, the mission may be too restrictive, and this becomes the eventual downfall of the company.

A broad mission statement can be just as dangerous. During the 1960s, many firms expanded because they thought their scope was too narrow, as was the case for many energy firms. These firms became involved in both allied businesses and unrelated businesses. However, in the 1970s and 1980s, many of these energy firms sold or tried to sell these businesses. They found that they did not have the expertise required to manage the new business or the understanding of the technologies and markets necessary to make the new business profitable. Having a broad mission statement can cause so much diversification that management will have little chance for control.

Because an organization's mission statement is the thread that binds its programs and activities, conscientious and continual reevaluation of the statement is necessary. In order for meaningful executive action to take place, the corporate mission must be clearly defined and communicated to all subordinates.

Once the mission statement has been established, the next step is to identify the corporate strengths and weaknesses. Some refer to this approach as a situation audit. This process consists of reviewing past performance and analyzing external forces, internal strengths, and internal weaknesses. Steiner [5] refers to this process as "WOTS-UP" analysis, which is an acronym for weaknesses, opportunities, threats, and strengths. Brainstorming and using an industrious research staff are two ways of developing a "WOTS-UP" analysis. However, most firms prefer less rigorous development techniques such as interviews with key managers or formal questionnaires aimed at discerning their thoughts or suggestions about opportunity, threats, strengths, or weaknesses.

The situation audit helps the company move toward their strengths and away from their weaknesses. Few companies study their strengths systematically and even fewer admit to or systematically study their weaknesses. Opportunities must be identified and exploited while threats must be delineated and kept at bay. Also, the skills and resources of the organization are assessed during the situation audit. Decisions are made concerning the commitment of these skills and resources to the various business opportunities.

In the formulation phase, the organizational objectives and goals are established. These objectives are dictated by the mission of the corporation and guided by the three basic objectives of all businesses: profitability, growth, and survival. All activity in the formulation phase is directed toward establishing the specific business or businesses in which the organization intends to engage — basic niche. The formulation phase gives the direction and the focus for the firm.

Implementation Phase

The implementation phase indicates the development of the operating strategy. Operating strategy flows from the basic strategy and delineates the second stage of executive action. This strategy defines the general and specific actions of the company. The implementation phase of the strategic design requires major effort on the CEO's part. The operating strategies must be numerous, and they must cover a variety of specific functions engaged in by the company. These strategies are then tested against the environment and other outside influences. Development strategies are part of the operating strategies, and they determine the organization's operating functions. Some of the strategies deal with items such as staffing requirements and others deal with long- and short-term planning requirements.

The operating strategies must be integrated with the financial strategies. The financial strategies deal with dividends, retained earnings, debt, cash flow, and budgeting. During 1982, the oil industry moved from debt financing to cash-flow financing. This turn of events completely altered the methods and means by which firms raise money to finance exploration budgets. These financial strategies determine the monetary resources that a company can commit and will definitely affect its risk orientation. The financial strategies are currently linked very closely to the basic objective of all organizations: survival.

Organization and Control Phase

After the strategic plan is organized in the formulation and implementation phases, the operating strategies are formulated and organized. The success of the company is often determined at this phase of the policy making. The allocation of skills and resources is used to affect the basic operating strategies.

The organization's objectives and goals are determined in the organization and control phase. The operational manager will be forced to live with these objectives and goals. These objectives may be stated in either quantitative or qualitative terms. However, objectives should not be time related and should be somewhat broad in their connotations. The objectives must be related to the strategy in such a manner that they serve to limit some of the choices available in the planning process.

In contrast to the objectives, the organizational goals are both specific and time based so that measurement of the organization's performance toward achieving the more general objectives can be made. These goals in many cases are program or project oriented. If the organizational goals are set only by corporate, nonoperating management, they can be unrealistic or even unattainable. Successful firms often use goal factoring, which is a negotiating process between operational management and corporate management, for determining realistic goals. Goal factoring will be discussed in greater detail in chapter 8.

Organizational restructuring is done at this point to make the organization more responsive to strategies, objectives, and goals. New courses of action may require a move toward centralization or decentralization. The purpose of restructuring is to make the structure of the organization compatible with strategy objectives and goals.

Formulation of guidelines for operational or developmental planning occurs in the organization and control phase. These formulations are essential for successful translation of strategies chosen in the implementation phase into series of goals and tasks. The completion of these goals and tasks can be measured by key factors that will be included in the management information system which monitors the efforts of the operational office.

Appraisal of the corporate culture should be done in the organization and control phase. If problems are evident, corrective action by the CEO should be undertaken to prevent encumberance of the operational plan. This insurance is important if the organization is anticipating a change in direction. "Corporate inertia" or the resistance to change in direction can deter achievement of the corporate objectives and goals in the strategic planning process.

The control system is designed in the organization and control phase. The primary purpose of a control system is to monitor the results of both tangible and intangible operational factors and to control future decisions. The management information system, that is, the feedback mechanism, is three pronged. One prong is the key factor reporting system that details progress made toward corporate goals and objectives and tracks the strategies for validity. Newman [6] breaks this portion of the control system into three more parts: (1) "steering controls," in which results are predicted and corrective actions are taken before the operation is completed; (2) "yes-no" controls or decision point oriented control (essentially a safety device) in which approval is required before the succeeding step is initiated; and (3) post-action control, or post audit, in which the action is evaluated after it is completed.

The two remaining components of management information systems deal with factors outside the organization. One component is involved with the operating intelligence system that compares the efforts of the company with those of competitors (competitor analysis). The last component entails scanning the environment to determine if the assumptions made concerning the operating environments were valid and, if not, to deduce what adjustments must be made.

Reformulation — Feedback and Review Phase

The reformulation — feedback and review phase is situated at the base of the strategic planning system. In this phase, management assesses its position and progress toward the corporate objectives and goals. This phase is dependent upon the control system and its ability to monitor activity and determine whether or not progress is being made. Further requirements of the control system are to determine problem areas and locate the source of the problems. The control system must also determine if there are fallacies in the strategies or planning assumptions.

Environmental scanning as part of the overall control system detects threat to the well-being of the firm. It also identifies new opportunities that were not originally assessed that can be incorporated into the basic strategies. Today, dynamic business conditions and the uncertain economic environment can render corporate strategies and plans obsolete. The reformulation phase will focus on one or more prior phases calling for changes in the operating strategies to deal with day-to-day changes in operating conditions. Organizational adjustments may be required to more efficiently execute the strategic plan. Also important, the design of the control system itself may require change to provide the vital information needed to effect the reformulation phase.

Adaptations to Meet the Needs of Diverse Businesses

Division and group management in larger, more diversified corporations needs flexibility to pursue its particular line of business. The strategic business unit, or SBU, which was pioneered at General Electric Company by McKinsey and Company, has become a popular approach. Under the SBU approach, operating strategies, goals, and objectives are handled at the group or division level. Ideally, the SBU gets emphasis from corporate management and the accompanying allocation of resources. Each SBU is in a particular line of business, which corporate management treats individually.

Corporate management can use a portfolio approach to analyze each of the various SBU businesses to make decisions and set strategies. One portfolio approach was developed in the late 1960s by the Boston Consulting Group (BCG). Their approach places business into a matrix, "cows" indicate businesses that are cash generators. "Dogs" indicate businesses that are losing money or have low growth potential. "Stars" indicate businesses with growth potential. "Question marks" indicate businesses that have high risk with doubtful potential. It is interesting to note that "question marks" were once known as "wildcats," a popular term for test wells. Depending on the category in which the SBU falls, the future operating strategies' objectives and goals are generated and superimposed on the operating levels. Figure 2.2 shows a stylized BCG matrix. Chapter 5 will show another portfolio approach that can be carried over into exploration planning.

This SBU approach has been emulated by a number of corporations with diverse operations. The management of each SBU, working with its staff, develops its unit's strategy. Top management then annually reviews and refines the SBU

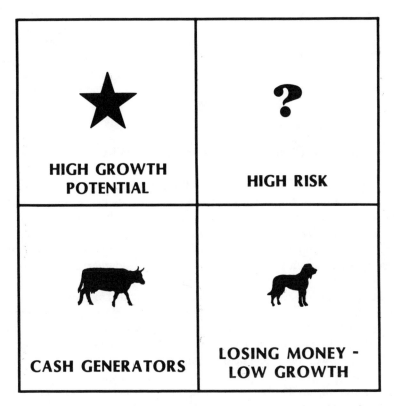

Figure 2.2. Stylized growth matrix based on Boston Consulting Group growth matrix.

strategy before approval. Once approved, SBU strategy is integrated with the corporate strategy and eventually becomes the basis for the SBU's annual budgets. The portfolio approach gives top management a method of analyzing both the SBU's past performance and its projected future performance.

The SBU provides the CEO with the means to determine the profitability of the various businesses. It provides the documentation needed to dispose of unprofitable businesses. This approach also provides the CEO with an understanding of the SBU strategies that can be used in discussion with the managers of the businesses. The SBU system has an advantage over mere budgeting or conventional long-range planning in that it forces upper management to focus on the competitive dynamics of the individual businesses. The system relates one business to another. According to Phillippe Haspaslagh of INSEAD, a European business school, half of the Fortune 1000 companies, by 1981, were using some sort of business matrix–portfolio approach in their planning. This approach, with its inherent complexity, provides a boon to the strategic planning consultants. The Boston Consulting Group, Bain and Co., Strategic Planning Associates, and Braxton Associates, were the four leading strategic planning firms. Other firms adopted the methodology, and strategic planning became the planning fad of the 1970s.

Where Strategic Planning Has Failed and Why

In the late 1950s and early 1960s the business atmosphere was stable and dominated by growth. Strategic planning was designed to take advantage of these trends and was very successful. However, uncertainties of all types came with the 1970s. The energy crisis caused by the Arab oil embargo had a major influence on industry. Petroleum-related industries experienced positive financial results because of this uncertainty; but for the manufacturing industries, the rise in energy costs was disastrous. Further uncertainty and complicated planning were caused by intensified government regulations. Strategic planning had to adapt to this rapidly changing environment, and this adaptation was not as rapid as management would have preferred. In fact, many strategic planning systems in use today are throwbacks to the 1960s, and the strategies and models in use do not have the flexibility required to deal with the changing environment.

The SBU and portfolio approach mentioned earlier was proposed to give the conglomerate company flexibility in its planning. Kiechel [7] points out the problems this approach is having in the market matrices. Many of the businesses have fallen into "dog" or "cow" categories, with market share and price being the major components in the matrix. The BCG type matrix does not give alternative actions and alternative opportunities. The current economic conditions have negated some of the effectiveness of the employed portfolio methods. A more important problem that Kiechel mentions is that the methodology which many consulting firms emphasize so strongly has not replaced thinking and decision making at the CEO level. Strategic planning failed to communicate to top management that the corporate strategy has to be implemented if it is to be useful. They could not merely follow the annual planning cycle that the consultants had devised. Corporations did not change the strategic planning process annually. The planning cycle became a routine exercise that did not aid key management in achieving corporate goals. This criticism is not directed at portfolio theory as method or as corporate strategy tool but rather at the "canned" planning methodology that was employed in the past by many planning consultants.

Accountability is less an issue when the economic atmosphere is good. However, in the mid 1970s when the business scene soured, accountability became an important issue in strategic planning. Control systems must be capable of defining problems before they become disastrous. As mentioned earlier, control systems have failed to give top management indications of strategies that were faulty or had become obsolete.

In the 1980s strategic planning will face another decade of rapid change and uncertainty. Strategic thinking will have to be innovative and farsighted to cope with these changes. Strategic planning systems will have to be redesigned and adapted to deal with these circumstances.

Making Strategic Planning Work

New directions are needed in the areas of accountability and flexibility for strategic planning to work. Accountability and flexibility can be divided into five essential subcategories: goal sharing, implementation controls, incentives, flexibility, and corporate culture.

Goal sharing is a necessary part of insuring accountability. The CEO as the chief planner is responsible for deciding the overall corporate goals and objectives. However, for planning to be successful, the line managers must be directly involved in the planning process within their domains, so that they can use their knowledge of specific areas of the business. Because only the line managers themselves are able to suggest operating goals that are realistic as well as challenging for their areas, they must help set their own goals and integrate them with the corporate goals.

Once the goals have been set, the next logical step is to find methods that can monitor progress toward these goals and objectives. Intermediate goals and milestones are best used for this purpose. The ideal control system should be capable of showing whether or not the strategy is being followed and is valid. This capability is essential to the iterative nature of strategic planning.

Roush and Ball [8] developed a valid concept for controlling strategies. Most of the control systems in the past were geared to performance criteria. Roush and Ball have proposed another control criteria, the critical implementation factor (CIF), that is keyed to monitoring the progress toward a desired strategy. The CIF measures the four critical implementation areas of organization, structure, culture, and control systems. These implementation factors serve the dual purpose of checking to see if the strategy is still valid and checking the performance criteria for validity. The CIF adds flexibility and aids contingency planning. The CIF should be used at higher levels of management, so to be effective, it should be controlled directly by the chief planner, the CEO.

Once controls are established, an incentive system must be tied directly to the control system so that the desired results are recognized. This system needs both a positive and negative aspect to make the system work.

Also important in the strategic planning system is flexibility. It is particularly necessary in the current operating environment because the ability to react quickly is essential. This can be accomplished through contingency planning and innovation. In contingency planning, outcomes are forecast for differing scenarios that are the results of changes in both the internal and external operating environment. Remaining flexibility must come through innovative and creative changes in strategic thinking, planning methods, and organization.

Contingency planning, which has become somewhat of a planning buzzword in recent years, is based on a set of scenarios, which have been hypothesized outcomes. Predetermined trigger points are used as decision criteria for scenarios on which contingency planning for the corporation is based. These may be significant changes in environment or variables, such as product prices, cost of capital, political climate, or internal factors, that have direct impact. Alternative strategic plans are then formulated for each contingent scenario.

Innovation, another aspect of flexibility, is more difficult to describe because it deals with creative processes. Creative processes must be matured within the corporate culture. "Corporate culture" refers to the mores and attitudes of the people in the corporate organization and determines how business is conducted. Innovation and creative planning may be hindered under these cultural constraints, and the oil business is notorious in this regard. Numerous case histories exist in which innovations failed because of the tendency to conduct business with the old-fashioned traditions and values of "oil men." Corporate culture as the primary topic has been discussed in numerous articles. Complete treatment of the subject has been given by Deal and Kennedy [9]. In their book, *Corporate Cultures*, the authors classify the types of culture and the effects of each. They make evident the fact that corporate culture can have either a positive or negative effect on operations and on the changes that management wants to enact through strategic planning. One of the main points that Deal and Kennedy stress is shared values between the company and its employees. A strong culture is described:

> *Sometimes the culture of an organization is very strong and cohesive; everyone knows the goals of the organization and these goals set the patterns for people's activities, opinions and actions* [9].

If there are shared values and employees know the reason for change, the chances of achieving the needed changes are greatly increased. Deal and Kennedy also propose methodology whereby cultures can be modified. In the situation early in the planning process, the CEO should analyze his corporate culture and set strategies to deal with it if it is a problem area.

Top Down and Bottom Up

All strategies, objectives, and goals have traditionally been passed down from the top to the operational people. This is done when the CEO knows intimately the details of the company operations. In large companies, the formulation phase, basic strategy, corporate mission, aims and goals, and the basic business objectives must come from the CEO and the corporate infrastructure. The implementation phase takes place largely in the corporate area. The primary problems with this sort of "top down" approach are that direction is not specific, and the objectives and goals are too general. The operating and business managers on the levels below need specific goals and objectives in order to get the proper direction at their level.

At the functional or operating level, the strategic objectives, goals, and strategies are often tangled by operational problems. If the goals and objectives are imposed from above upon the operating level, then these line people will have difficulty integrating planning into their daily operations. The operational manager's vision of decision making is on a day-to-day basis. Often the long-term goals and objectives are lost to the operational manager because, at this level, a manager's activities are most often measured by short-term results, not long-term criteria, causing operational managers to shun long-term strategies, objectives, and goals. Perhaps as much as any industry, petroleum exploration must rely on long-term strategies because significant discoveries may be years apart for many companies. So that strategic goals can be accomplished, objectives met, and strategies followed, it is necessary to design a hierarchy of decision making that is tied to criteria for both long- and short-term goals. In other words, operational planning must tie directly to strategic planning; this can be done using control and incentive systems.

Goal acceptance is dependent on participation at all levels. This participation may be difficult when superiors do not rely on their subordinates. Compounding the problem, subordinates may sometimes provide invalid data to guarantee their performance. If participation in goal setting is not feasible, then goals must be clear, reasonable, and meaningful, and rewards must be directly and openly tied to performance. If individuals have participated in setting challenging goals, then they will be motivated to achieve those goals. Acceptance, motivation, and achievement will be greater if there is strong intercommitment to the mutual goals.

References

1. Ewing, D.W.: "Corporate Planning at a Crossroads," *Harvard Business Review* (July–Aug. 1967) 45, no. 4, 77–86. Last republished in *Long Range Planning for Management*, Harper & Row, New York (1972).

2. *Webster's New International Dictionary of the English Language*, second edition, unabridged, G. & C. Merriam Co., Springfield, MA (1959).
3. McNichols, T.J.: *Executive Policy and Strategic Planning*, McGraw-Hill, New York (1977).
4. Anthony, R.N., Dearden, J., and Vancil, R.F.: *Management Control Systems*, Richard D. Irwin, Homewood, IL (1980).
5. Steiner, G.A.: *Strategic Managerial Planning*, Planning Executives Institute, Oxford, OH (1977).
6. Newman, W.H.: *Constructive Control — Design and Use of Control Systems*, Prentice Hall Inc., Englewood Cliffs, NJ, (1975).
7. Kiechel W.: "Corporate Strategists Under Fire," *Fortune* (Dec. 17, 1982) 106, no. 13, 34–39.
8. Roush, C.H. and Ball, B.C., Jr.: "Controlling the Implementation of Strategy," *Managerial Planning* (Nov.-Dec., 1980) 29, no. 4, 3–12.
9. Deal, T.E. and Kennedy, A.A.: *Corporate Cultures — The Rites and Rituals of Corporate Life*, Addison-Wesley, Reading, MA (1982).

Further Readings

Ackoff, R.L. and Emery, F.E.: *On Purposeful Systems*, Aldine-Atherton, Chicago (1972).
Ackoff, R.L.: "Management Misinformation Systems," *Management Science* (Dec. 1967) 14, no. 4, 147–156.
Andrews, K.R.: *The Concept of Corporate Strategy*, Dow Jones-Irwin, Homewood, IL (1971).
Ansoff, H.I.: "The State of Practice In Planning Systems," *Sloan Management Review* (Winter 1977) 18, no. 2, 1–24.
Breech, E.R.: "Planning the Basic Strategy of a Large Business," *Planning the Future Strategy of Your Business*, (E.C. Bursk and D.H. Fenn, Jr. eds.), McGraw-Hill, New York (1956).
Cleland, D.I. and King, W.R.: *Management: A Systems Approach*, McGraw-Hill, New York (1972).
Drucker, P.F.: *The Age of Discontinuity*, Harper & Row, New York (1978).
Haines, W.R.: "Corporate Planning and Management by Objectives," *Long Range Planning* (Aug. 1977) 10, no. 4, 13–20.
Harrison, F.L.: "How Corporate Planning Responds to Uncertainty," *Long Range Planning* (Apr. 1976) 9, no. 2, 88–93.
King, W.R. and Epstein, B.: "Assessing the Value of Information," *Management Datamatics* (Sept. 1976) 5, no. 4, 171–180.
King, W.R. and Cleland, D.I.: *Strategic Planning and Policy*, Van Nostrand Reinhold Co., New York (1978).
Lorange, P. and Vancil, R.F.: "How to Design a Strategic Planning System," *Harvard Business Review* (Sept. 1976) 54, no. 5, 75–81.
Stevenson, H.H.: "Defining Corporate Strengths and Weaknesses," *Sloan Management Review* (Spring 1976) 17, no. 3, 51–68.
Vancil, R.F.: "Strategy Formulation in Complex Organizations," *Sloan Management Review* (Winter 1976) 17, no. 2, 1–18.

3

Traditional Exploration Strategies

What are exploration strategies? Exploration strategy is a much used term that is seldom defined and has a different connotation at different levels within an exploration operation. In a literature search, articles concerning exploration strategy did not address the strategy itself but addressed the subsequent actions within the exploration process.

The View of Individual Explorationists

Individual explorationists probably have the most simplistic view of exploration strategies. They are likely involved with three or four plays at a time, and their exploration strategies must address questions such as: How did their companies get into this play? Or how does a specific prospect being generated fit with what other companies are doing in this area? Strategies are limited by the environment of the play. Questions concerning the play are: Is there open acreage? Do deals have to be made for the acreage? The explorationists know that there is competition for capital and that their plays and prospects will be compared with those of other explorationists in the company.

Unless the organization is small, individual explorationists may have trouble relating their views on exploration strategies to the views of the organization. If the organization is large, individual explorationists may even be unsure of the overall direction of a particular office. If explorationists are at the level of district geologist or geophysicist, they may have a somewhat larger perspective but still a limited view of the total organizational exploration strategies. Professionals at this level are basically technical people and unconcerned with business goals, objectives, and strategies. There is also a cultural component based on regional loyalties to their exploration strategies. Individual explorationists are inclined to be biased toward areas where they are personally involved and have had experience, which, in turn, influences their exploration strategies.

The View of Exploration Managers

Exploration managers have a broader view of exploration than their subordinates. Their exploration strategies should be answering these questions: Should we get into this play? When? And more importantly, when should we get out of this play? Exploration managers should be aware of the environmental factors that govern the above questions.

Exploration managers are more concerned with risk than their subordinates. Consequently, they must decide how to spread risk in their exploration program. Exploration managers know the physical limitations and constraints within their spheres of influence. Their strategies must deal with components of their exploration environment. Probably the foremost component is their capital expenditure, or their budget. Strategies must include measures to handle exposure of budget funds in relation to opportunities. In other words, exploration managers must figure out how to get the "biggest bang for their buck." Their experience and background will strongly temper their strategies. Those managers who have thorough knowledge of their particular areas of responsibility will be able to determine realistic risk intuitively. In addition, this experience will allow them to judge available potential in an area. This intuitive approach to assessing risk and potential can be used very successfully if the exploration manager has the necessary experience and judgment. However, there are few managers with this talent.

In short, managers have a wider view than their subordinates of the organization's business goals and objectives, but they generally do not have enough specific information to completely coordinate exploration strategies with the organization's business strategies, goals, and objectives.

The View of Land Managers

Land managers in some offices are also directly involved in formulation and implementation of exploration strategies. If land managers are given the latitude, they may add a dimension to the company's exploration strategies that are not in the explorationist's normal domain. They may make acreage plays in potential areas, but they usually act with some guidance from the explorationists. It is essential that land managers stay ahead of industry by buying inexpensive acreage in areas before industry activity raises lease costs. If a company has a good land position, it will allow the land manager to use the tools of the trade, such as farmout options and contributions, to test the acreage without the high cost of test well drilling. In competitive areas, the acreage position can make or break an exploration program. If a company has to drill to earn an acreage position, it could lose its competitive position. Being forced to pay a disproportionate share of test well costs, as would be the case in the usual farmin, will definitely increase finding costs.

The View of Operational Managers

Operational management, to whom the exploration manager reports, has a view of exploration strategies different from that of the exploration manager or land manager. The operational manager lives in a far more complex world. It is at this level that exploration strategies interface organizational business strategies. Operational managers view exploration strategies from a broad overall perspective but still must contend with physical constraints. Risk concerns them in an overall sense but not on a prospect-by-prospect basis. Operational managers have to balance available capital and use methods to control risk. In some companies, the operational manager, in order to survive, must be concerned with short-term results that are contrary to the long-term goals and objectives of the organization. This frequently forces compromise that is not completely within the overall interests of the company but will allow the operational manager to survive.

With regard to the exploration budget, operational managers must be selective in their exploration strategies to maximize capital allocation among various exploration expenditures, such as lease purchases, geophysical costs, and drilling contributions. By doing so, they may be able to select strategies that will be more economically effective than straight test well drilling strategies.

In larger corporations, there will be several layers of operational management. The vice-president in charge of the exploration function is the main interface between the organization's business strategies and the overall exploration strategies. At this level of the organization, exploration and production operations compete with other types of business opportunities, especially in companies that have diverse operations, such as in integrated major oil companies.

The View of International Explorationists

Only domestic exploration strategies have been discussed thus far. Foreign operations often do not have as many levels of exploration strategies because they operate differently than domestic operations. In foreign operations, concessions that are given to large international companies consist of millions of acres. The detailed strategies needed in the domestic areas do not apply in those cases. In most cases, the strategies were determined before the acquisition of the concession. The main formulations of politically risky exploration strategies are made early, when decisions about which countries or even continents the company wants to explore are made. The next level of strategy is tailored to meet requirements of the concesson grantors.

The costs and risks incurred in foreign exploration operations are considerably higher than in domestic exploration operations. The rewards also should be higher. If the organization participating in a foreign concession cannot afford to stay until success is experienced, then other measures and strategies will be needed to spread the risk. This requires that the company be aware of its total capital exposure. The large capital commitment that is required to engage in foreign exploration operations requires that exploration strategies directly interface with the organization's business strategies, so that there will be fewer problems with operational management's understanding of the organization's goals, objectives and strategies.

Summary

The traditional exploration strategies have been portrayed in this chapter as they are currently handled by most companies. Exploration strategies are viewed differently by the individual explorationist, exploration manager, land manager, operational manager, and international explorationist. In the following chapters, environmental analysis and the strategic planning process will be linked to specific exploration strategies so that they will interface with the organization's business strategies.

4

Examining the Environment

Environmental Analysis

As was discussed in chapter 2, one of the first components in strategic analysis is understanding the environment in which exploratory operations for the firm will take place. Strategic planning requires that this environment be properly described. If the environment is not properly understood, poor decisions or even grave errors may result.

The environmental analysis can be broken down into several segments addressing different characteristics of the exploration environment. Two major characteristics are the physical maturity or exploration history of the basin or play being explored and the associated forecast of future field sizes. Another characteristic that should be considered is social environment, often important in many of the areas where petroleum exploration firms operate. Also, available company resources or the inventory of strengths and weaknesses derived from the situation audit give internal characteristics to the overall exploration environment. Such an inventory enables exploration management to appraise its staff's knowledge and experience and fit them to the geological environment.

As will be seen later in the chapter, after the environment has been analyzed for physical, social, and internal characteristics, the next strategic consideration is the analysis of competing firms. Most of the sources of data and methods used for environmental analysis are closely related to those used for competitor analysis, so both are covered in this chapter.

Basin or Play Analysis

The historical basin analysis phase of the environmental review can be accomplished with data from a variety of sources, from periodical literature and company records to established commercially available computer data bases. Using commercially available data bases is usually the easiest and most effective method of obtaining the needed basin or play information.

The historical reserves distribution is probably the most important of environmental variables but is generally the hardest to evaluate. However, there are various ways to estimate field sizes from other available data. Field reserve sizes can be readily estimated using production performance techniques to estimate ultimate yields from commercially available production history.

Figure 4.1 summarizes schematically some of the key environmental variables that should be examined in the basin analysis or physical phase of the environmental review. Field and pool data from commercial data base systems such as the Petroleum Data System (P.D.S.) are combined with other sources of reserves information (such as estimates of ultimate production made from production history) and sorted by geological and historical limiting factors (such as discovery date, age and formation, trap type, etc.) to give a basin reserves analysis.

Figure 4.1 shows what conditional reserves distributions might look like. The first, larger plot is the total historical discovery size distribution for the basin or play. The smaller plots give the discovery size distribution broken down by important environmental variables. Each plot is a histogram of discovery size for all fields applying to a particular value (or range) of the environmental attribute listed below the plot.

The first set of small plots in figure 4.1 gives the field discovery size distributions for ten-year periods in the basin's exploration history to give a general time framework. The sets of plots below the time plots show the reserves distributions for different trap types, finding methods, producing age, lithology, and producing depth.

The lithology and geologic age of the producing fields and reserves can help delineate the geological history of the accumulation of hydrocarbons and should indicate trends that might continue into the future. The trap types and discovery methods are related and give insight into various periods in the basin or play exploration history. Many of the environmental variables besides reserves can also be statistically related to each other in a manner similar to the plots in figure 4.1. For example, if discovery depth is compared to discovery year, a definite deepening trend will almost invariably result because shallow fields are found early in a basin's history. However, as technology develops and geological "keys" are found, some of the trends may reverse themselves temporarily.

In addition to the above geological variables, exploration activity should also be included in the environmental analysis. Industry activity information is usually taken from well data bases as well as from field data sources. The number of test wells drilled per year generally gives a good indication of the level of industry interest in the basin over time. Also, the number of discoveries divided by the

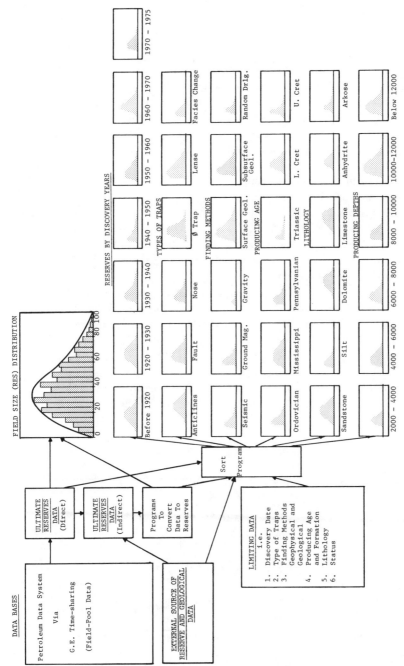

Figure 4.1. Flowsheet showing conditional reserve distributions for various environmental factors by geologic play or basin.

27

number of new field test wells (indicated by Lahee classification) drilled gives estimates of industry chance of success in the basin over time. Drilling densities can be calculated by dividing the number of wells drilled in each play by the total area of the play and compared to environmental attributes such as success ratio. The drilling density data can also be used to compare plays within the basin or, on a larger scale, to compare basins. Also, some of the commercial data bases include a well cost-to-depth analogy that can give relative costs of exploratory drilling in recent years.

If facilities within the exploration firm do not include computer systems to perform the statistical and data handling methods required for the basin analysis, "canned" or commercially available systems such as the Statistical Analysis System (SAS) and the Statistical Package for Social Sciences (SPSS) can be used. The SAS and SPSS packages also include graphical capabilities for producing plots and graphs of environmental relationships like those discussed above. Example plots produced with SAS are shown in figures 4.2 and 4.3

Figure 4.2 shows total reserves (in thousands of equivalent barrels) found throughout the Big Horn basin separated by geological formation and depth. The plot summarizes several characteristics of the Big Horn basin in an immediately accessible format. Indicated by figure 4.2, the largest producing formations in the Big Horn basin are Phosphoria, Tensleep, and Madison. The largest proportion of oil is contained in the Tensleep formation in shallow depths. In the deepest depth zone (deeper than 8,000 feet), more oil is contained in Phosphoria than in Tensleep. The actual reserves figures for each division are also included in figure 4.2.

BLOCK CHART OF SUMS

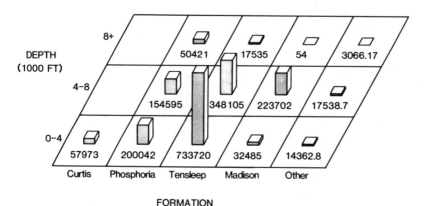

Figure 4.2.   Total reserves in thousand barrels found in the Big Horn basin by geologic formations and depth (SAS graph format).

BLOCK CHART OF SUMS

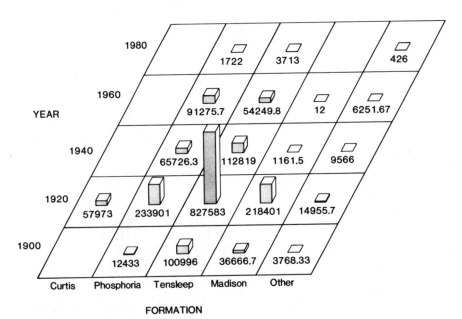

Figure 4.3. Total reserves in thousand barrels by geological formation and discovery year in the Big Horn basin (SAS graph format).

Figure 4.3 adds historical perspective to reserve distributions in the Big Horn basin. Total reserves in thousands of barrels of equivalent oil are separated by geological formation and discovery year. From this plot, the general decline in total oil found per year is evident for all three major formations. However, displaying the number of test wells drilled per year shows that industry activity in the basin has increased greatly (fig. 4.4). In the 1920s and 1930s (when the largest proportion of oil was discovered), exploration activity was only a fraction of what it would become in the 1960s and 1970s. The exploratory maturity of a basin or play is more accurately measured in number of exploratory wells than in years. The reason for this should be evident from figure 4.4 in which the number of wildcat wells drilled in the Big Horn basin in an average year increased tenfold between 1940 and 1955. Figure 4.5 shows the decline in average field size versus the number of new field wildcats drilled in the Big Horn basin. Between 1914 and 1980 there were approximately 1100 new field wildcats (test wells) drilled that found 140 fields. The average field size for each data point in figure 4.5 is the mean of ultimate reserve productions in thousands of equivalent barrels in a particular two-year

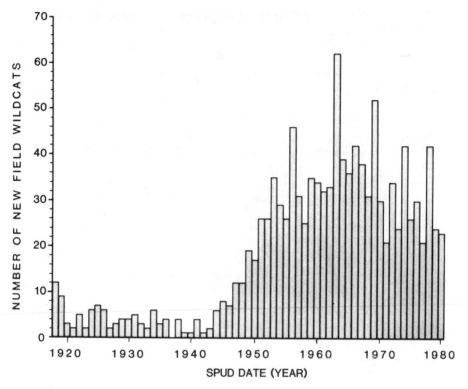

Figure 4.4.   Wildcat wells drilled in Big Horn basin from 1918 to 1980. Note increase in drilling 1940 through 1955.

period. The years were then converted to cumulative number of wells drilled up to the end of each of the two-year periods. For reference, years have been super-imposed on the cumulative well axis to give some indication of the basin's history. Each point represents the expected discovery size at a single "frozen" point in the exploration process and does not represent the average for all previous years. This method and the relatively small sample size causes some variation in the data points. The straight line in figure 4.5 represents a least squares fit through the data points. Considering that the discovery size axis is on a logarithmic scale, the decline in average discovery size is quite pronounced.

Similar plots were made for discovery sizes in the Permian basin using data gathered by the Dallas Field Office of the Energy Information Administration of the Department of Energy. The Permian basin of west Texas and southeast New Mexico is a very mature basin containing many geologically diverse plays. Figure 4.6 shows the total number of discoveries versus discovery size for all fields

BIG HORN BASIN
DISCOVERY SIZE vs. WILDCATS DRILLED

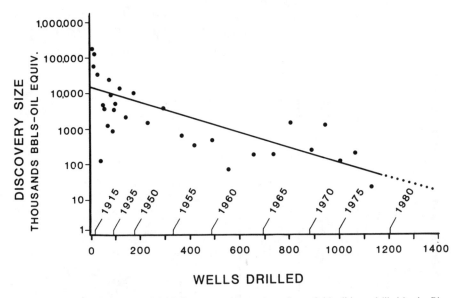

Figure 4.5. Decline in average field size versus the number of new field wildcats drilled in the Big Horn basin from 1914 through 1980. 1,100 wildcats were drilled in this period.

discovered between 1910 and 1974 in the Permian basin. Figure 4.6 demonstrates the skewed or lognormal distribution of field sizes occurring in nature where larger fields become increasingly rarer. Note that 4494 fields were recorded that had less than one million equivalent barrels of oil while only about 500 fields were reported with reserves between one and two million barrels. The field sizes double for each of the bars on figure 4.6, so the few fields falling on the right end of the scale are extremely large, some exceeding 500 million equivalent barrels. Figure 4.7 is identical to figure 4.6 except that only those fields discovered between 1963 and 1974 were included.

The average reserves discovered per test well for all test wells drilled in a basin gives a rough indication of how lucrative the basin is. Figure 4.8 shows the average reserves discovered per test well for the Permian basin from 1910 through 1980. Data was only available through 1974 so the last bar on figure 4.8 is partially extrapolated. Some well data before 1910 is probably missing so the first bar may be slightly overstated, but the decline in average reserve additions per exploratory well is still obvious. Average reserves per test well declined from at least five million equivalent barrels per well to about two-hundred thousand equivalent barrels per

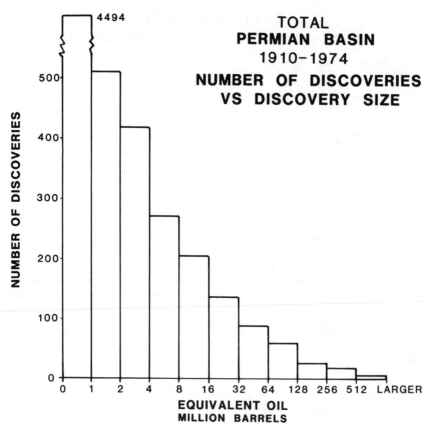

Figure 4.6.   Skewed or lognormal distribution of fields found in the Permian basin from 1910–1974.

well. A graph like figure 4.8 can be very useful for performing test well economics because a good estimate of expected test well reserve additions can be derived. The decline in average reserve addition means that a straight historical average based on data that are several years old may be overstated—hence, even the latest average, for 1970 to 1974 in figure 4.8, may be too large to use today.

The example illustrations given so far have been for two mature but very different basins. The Permian basin covers a large geographical area and contains over 6000 producing fields. The Big Horn basin is smaller with less than 300 producing fields. The Permian basin is a complex of smaller basins and shelf areas while the Big Horn basin is a single, deep intermountain basin. Production for the Permian basin is fairly evenly distributed over the entire basin with some local

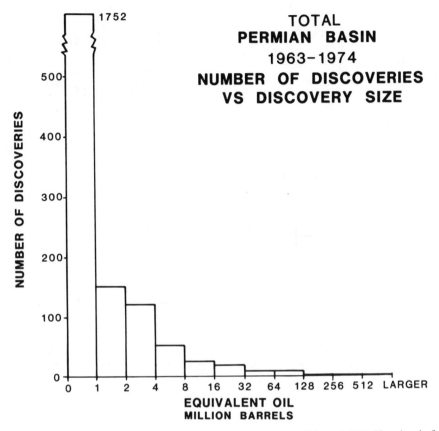

Figure 4.7.   Distribution of fields found in Permian basin from 1963 through 1974. Note that the field sizes found compared with figure 4.6 demonstrates that large fields are found early.

concentrations. In the Big Horn basin, production is concentrated in the anticlinal features in the Paleozoic beds on the periphery of the basin while the center of the basin is almost completely untested. The basin analysis phase of the environmental review is used to help identify differences in prospective areas, like the differences between the Permian and Big Horn basins.

As an example of how activity levels can be related to exploration results for different prospective areas, figure 4.9 compares drilling density to success ratio for 46 AAPG/CSD signified basins, all based on well data taken from the Petroleum Information Corporation (P.I.) data base for wells drilled between 1975 and 1981. The trend line indicated on figure 4.9 was shown to have no statistical significance.

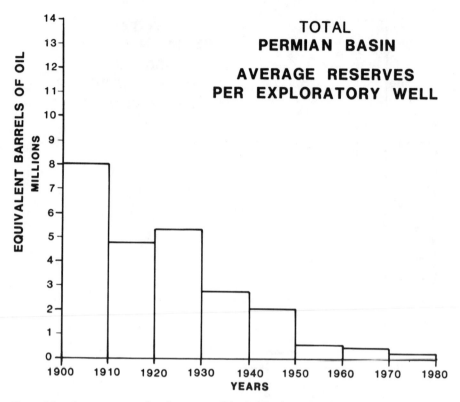

Figure 4.8.   Average reserves found per test well by decades from 1910 through 1980 (data for last decade are only through 1974).

That is, no solid relationship was implied between denser drilling (which should normally imply tighter well control) and higher success ratios. Hence this nonintuitive result was derived through environmental analysis.

Data Sources

The historical data that are required for basin analysis can be used for the forecasting techniques discussed in chapter 6. There are a number of digital computer data bases that can be used for basin analysis. The Petroleum Data System (P.D.S.) is affiliated with the University of Oklahoma in Norman, Oklahoma. P.D.S. maintains a series of data bases for oil and gas fields in the continental United States and Canada. Appendix A was prepared by P.D.S. personnel and gives more detail on their available data and some approaches to retrieval of data.

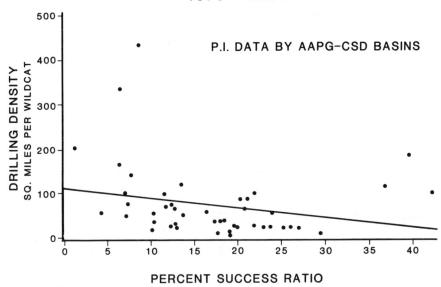

Figure 4.9. Drilling density compared to success ratio for 46 AAPG–CSD basins for wells drilled 1975 through 1981.

DATA SOURCE: Petroleum Information Corporation.

Petroleum Information Corporation (P.I.) has a number of proprietary data bases. P.I. has, independently from this text, derived methods for systematically accessing their various commercial data bases in a logical sequence to give a synergistic approach to basin analysis. Appendix B has been prepared by P.I. to detail their data and methods.

Production data is available on P.I. and Dwight's Energy Data Inc. Production data can be used to refine field and pool data. Also, time series production information can be combined with production decline curve fitting techniques to estimate production or field sizes.

Since the Department of Energy assumed the estimation of hydrocarbon reserves from the American Petroleum Institute (API) and American Gas Association (AGA), they have been publishing yearbooks and reserves summaries similar to those done earlier by API–AGA. Although many proprietary data are not currently available to the oil industry, plans exist to include more industry reported data in publicly accessible data bases.

Social Risk Assessment

Very little work has been done on the sociological aspects of environmental analysis for petroleum exploration. Investigation of sociological influences is warranted when a company has plans to invest millions of dollars in a new area. Kent and Taylor [1] of the Foundation for Urban and Neighborhood Development (FUND) of Denver, Colorado, discusses how sociological influences can add greatly to the cost of large scale projects. Kent gives an approach to social risk management that is in three basic parts: (1) social resource mapping to categorize prospective areas by social risk, (2) situational assessment of specific sites, and (3) social impact management, which involves cooperation between citizens, government, and industry to achieve common goals. Social risk management is designed to minimize delays and expenses due to mitigations between industry and local citizens or government. FUND has prepared data relating to social risk assessment that appears in Appendix C.

Situation Audit

The environmental characteristics examined so far are external to the exploration firm, that is, they would be the same for any firm. However, the internal characteristics of the firm contribute to the total exploration environment. A situation audit is performed to assess these internal characteristics. The situation audit attempts to identify the company's strengths and weaknesses.

The situation audit should begin by reviewing the company's past performance by play. Those plays in which the company has had extraordinary success should be identified and the reason for greater success analyzed. The success may be attributed to some special seismic interpretation based on proprietary modeling methods, for example. Or perhaps a strong acreage position, achieved by an aggressive land department, established a dominant position in the play. Whatever the reasons for success, an attempt should be made to identify advantages so that they can be nurtured within the play and applied to similar plays.

The performance review should also focus on plays where results have been poor, since the situation audit also identifies weaknesses within the company. Reasons for the weaknesses should also be found. Was the failure due to ordinary bad luck? Did a particular exploration program include sufficient wells to draw conclusions about its effectiveness in the future? After all, only a very small percentage of exploratory wells make significant discoveries in any play. Weaknesses may also be hard to define because of the human tendency to refuse admitting failure.

Besides the review of operational performance, the situation audit must also examine financial and human resource aspects of exploration. Is the firm's financial backing complete enough to increase operations in those plays where success has been achieved? Is the firm's financial position strong enough to mount an exploration program in an unproved frontier area? Can geologists with the proper background and experience be employed in the new frontier area? Will drilling rigs be available?

The situation audit is no more than an inventory of the firm's strengths and weaknesses. The strengths will help give physical shape to corporate strategic goals. The weaknesses will hopefully identify problem areas and potential snags before they become all-consuming. Methods for managing risk and setting exploration strategies in specifically translatable forms will be developed in chapters 5 and 6.

Competitor Analysis

Much can be gained for strategic planning by examining the success of competing companies. Just as the situation audit identified strengths within the firm, competitor analysis can identify strengths in competing firms for use in strategic planning.

There are a number of publicly available data sources for competitor analysis. The same computer data bases used for the environmental analysis above are of use in performing competitor analyses. The P.I. DAAS system can be used to identify well operators within a play and the cost analogy can then be used to determine how much is being expended within the play by competing firms. Using the same well data together with the Lahee well classification, success ratios can be determined for competing firms and used to identify a particular firm's abnormal success. Field data from the P.D.S. data bases can then be used to qualify these successes by field size. P.D.S. includes a data base of AAPG/CSD records by year identifying field size by Lahee field classification. Currently a project is under way to update these field size categories with follow-ups three years after discovery and again six years after discovery.

Beginning in 1978, the Securities and Exchange Commission (S.E.C.) has required publicly owned companies involved in oil and gas exploration to make extensive disclosures of current reserve quantities, changes in reserve quantities, and costs incurred as part of exploration. These disclosures, known as 10-K reports, are a valuable and easily obtained source of data for competitor analysis. The 10-K reports should be used with caution, however, by someone familiar with S.E.C. regulation of data reporting. Data in the 10-K reports are consolidated for whole companies but can be combined with information from computer data bases like P.I. and P.D.S. to divide reserves by local operations.

Financial data on the 10-K reports include data to determine potential future value of stock, net income, and operating costs by category. Reserves data include reserves at year end and reserves added by acquisition, by improved recovery, and by discoveries and extensions. The S.E.C. requires separate reporting of developed and undeveloped reserves with associated value projections. Net well information is of interest to explorationists and includes productive and dry net wells for both exploratory and development wells for the United States, Canada, and foreign countries. Acreage holdings for the United States, Canada, and foreign countries are also reported.

Competitors can be compared according to the increases they have made in their reserve bases. The effectiveness of the operating expenditures can also be examined by using cost data. Reported costs include exploration, development, lifting, property acquisition costs, and book values.

Arthur Anderson & Co. [2] published a survey of 10-K report data through 1981 for 234 exploration companies. They grouped companies into categories for major companies (16 were included), independent companies (107), integrated companies (24), pipeline/utility companies (28), and diversified companies (59). Arthur Anderson & Co.'s definitions for the categories were:

> *Majors* — *integrated companies having more than one-billion equivalent barrels of proved reserves.*
> *Independents* — *companies involved principally in exploration and production activities.*
> *Integrated* — *integrated producers other than the majors.*
> *Pipeline/utility* — *companies that are principally involved in rate-regulated pipeline or utility operations.*
> *Diversified* — *companies with significant operations in other industries.*

The survey completed by Arthur Anderson & Co. did not attempt to analyze all 10-K data. Its emphasis was primarily on financial and reserve data. However, the survey covers many important elements of competitor analysis and provides an excellent starting point for the exploration firm wishing to perform competitor analysis. Hopefully, 10-K data will eventually be made available through some computer accessible data base.

Summary of Environmental Analysis

The environmental analysis should be concerned with three important segments addressing physical (or geological), sociological, and internal characteristics. Of these characteristics, physical attributes are the most important and are detailed by basin or play analysis. Considerations internal to the exploration firm are identified in a situation audit. Competitor analysis is closely related to environmental analysis and provides further input to strategic planning.

The information gathered by the environmental analysis is used in chapters 5 and 6 to manage risk and to evaluate and derive specific exploration strategies.

References

1. Kent, J.H. and Taylor, D.C.: "Social Risk Management — A Successful Project Development Approach," *Military Engineer* (May–June 1983) 75 no. 487, 268–71.
2. Arthur Anderson & Co.: *Survey of Oil and Gas Disclosures*, Arthur Anderson & Co., Houston, TX (1982).

5

Strategic Analysis: Risk Methods

Incompleteness of Expected Value Analysis

The most commonly used methods of project evaluation in petroleum exploration rely on expected value analysis as criteria for undertaking the project. Expected value methods summarize the attributes of a project by calculating the average outcome on which decisions are based. But to assume that the average case can be "expected" in petroleum exploration is pure folly except in the extreme long run.

Consider, as a simplistic example, the connotation of so elemental a measure as chance of test well success, one of the primary attributes on which most expected value calculations are based. Suppose the chance of success is 10%. Obviously, we do not expect 10% of the well to be successful: we expect either a discovery or a dry hole. Does the chance of success mean that if we drill in the same well location over and over again we would observe success 10% of the time? Obviously not, if the first well was successful, they all would be successful (ignoring the possibility of incorrectly testing the discovery, which is not measured by the chance of success). Does the chance of success mean that if we drill a great many identical prospects then 10% of them would be discoveries? First of all there are no identical prospects. Second, if the prospects were identical then they would yield identical results: that is, all discoveries or all dry holes. Does the chance of success mean that if a great many structurally identical anomalies existed in the same general area then 10% of them would have trapped migrating hydrocarbons? Perhaps this is getting conceptually closer, but still there are problems. The chance of success was not based on our perception of the anomaly (which is rather sketchy above ground) but on results from previous wells — wells drilled on unidentical prospects. Does the chance

Portions of chapter 5 originally appeared in altered form in SPE Proceedings; copyright 1983, SPE-AIME (Quick, A.N. and Buck, N.A.: "Portfolio Analysis of Exploration Strategies," paper 11300, Proc., SPE Hydrocarbon Economics and Evaluation Symposium, Dallas, TX 1983).

of success mean that if a great many anomalies in the general area were delineated as clearly as possible above ground and the most promising of these were drilled then about 10% of them would be discoveries? This may be getting close but the imprecision in the estimate is now becoming apparent. Probably all the chance of success tells us when applied to the next test well is that we are fairly sure (about 90% confident) it will be a dry hole! All this is not meant to confound the reader but to show that petroleum exploration should not be evaluated strictly in terms of single prospects or expected outcomes but in terms of programs designed to keep risk at bay while making the desired additions to company assets. The evaluation of a multiwell program or exploration strategy calls for the assessment and setting of risk levels. The following section on viability analysis is concerned with the assessment of program risk. The section that follows it on portfolio analysis makes use of the assessment methods to control program risk in choosing an exploration strategy.

Viability Analysis

The most obvious goal of any multiwell strategy is program viability or survival. A viable program is one that sustains its own needs for capital. A program that has had too many dry holes in a row loses its viability either because no more capital is available for additional wells because exploration management no longer feels that the program should be pursued (perhaps the estimates of potential were overstated). In the latter case, all is as it should be because there has been a learning process. But the first case, in which a program must be abandoned because of adverse chance outcome, can have disastrous effects on the exploration firm and in the case of smaller companies even bankruptcy. This probability of not meeting a minimum goal for survival is often called "gambler's ruin," and methods exist for managing it.

Gambler's ruin need not necessarily imply total bankruptcy and ruin of the company. Although it can be used to analyze the probability of ruin for smaller firms or for large high-risk exploration programs, in a medium to large firm, gambler's ruin is more useful for analyzing the possibility of achieving a particular objective. For example, exploration management may have stated as a goal finding a 100 million barrel field during the course of a five-year drilling program. The gambler's ruin approach can be used intact where "ruin" means no such field was found. Or in another case, management may want an expensive international drilling program to maintain a certain level of self-sufficiency to minimize the need to draw on "bread and butter" domestic production over the next few years so that growth is not jeopardized. In this case, "ruin" means that the international program did not demonstrate the desired success over the short term. Because of these dominant related applications and because the term gambler's ruin carries a certain stigma (people often refuse to consider the case of true ruin) perhaps a better term for similar studies would be "viability analysis" or "measures of self-sufficiency" because ruin and cataclysmic collapse of the firm are rarely involved.

Viability analysis can be used at all levels in the firm to integrate localized and corporate exploration goals. At its basic level, viability for a multiwell program can be determined. However, the same concepts and methods can be applied to studying the viability of a corporate exploration strategy made up of several multiwell programs each in its own play. In this way, the probabilities of goal achievement, whether in a single play, large province, or worldwide, can be calculated. This serves the dual purpose of helping to keep strategic goals realistic at all levels. If local management can show corporate management what chance they have of meeting their own goals, corporate management is better able to set local goals that properly integrate with company-wide goals in maintaining the overall viability of the firm. As will be seen, these viability measures can be combined into "portfolios" for maximizing return while managing risk.

Probability of No Successes

If the probability that a test well will discover a new field is p, then the chance that it will be a dry hole is $(1 - p)$. If several wells have probability of success of p_1, p_2, p_3, etc., then the probability that one of them will yield successes (shown as v) is

$$v = (1 - p_1)(1 - p_2)(1 - p_3)\ldots \quad (5.1)$$

If the chances of success for all wells are taken as equal (such as with new prospects in the same play), designated p, and there are n wells, then equation (5.1) can be condensed to

$$v = (1 - p)^n. \quad (5.2)$$

Let the exploration cost of a test well be x. If capital available for test wells is C, then the number of test wells that can be drilled before the remaining capital drops below x is $[C/x]$. (The brackets around the ratio indicate the "floor" or largest whole integer contained in C/x.) Equation (5.2) can be written

$$v = (1 - p)^{[C/x]}. \quad (5.3)$$

This equation gives the probability, v, that the initial capital will be exhausted by a succession of dry holes. Arps and Arps [1] gave this expression in their discussion of gambler's ruin.

For example, suppose the test well budget has been set at \$10 million and test wells cost \$1 million each with chance of success of 15%. If working interest is to be

100%, then $C = 10$ (dropping the million), $x = 1$, $p = 0.15$, and the resulting probability that the initial capital will be consumed without a success, v, is about 0.20 from equation (5.3). Obviously, if the working interest is reduced and more than 10 wells are participated in, then the chance of program failure will be reduced. Suppose that the working interest is reduced to 50%. Intuitively, it may seem that the chance of program failure might be halved also. However, now, $x = \frac{1}{2}$, 20 wells can be drilled and $v = 0.039$. In other words, halving the working interest reduced the probability of program failure from 20% to 4%.

If the initial capital divided by the test well cost is equal to the number of wells to be drilled, then equation (5.3) can be used to analyze the viability of an n well program. (In this case, revenue from successes is not required for continuation of the program as originally planned.) Also, equation (5.3) can be used if n is not large enough to exhaust both the initial capital and the return from one success. For the example given above where C/x was 10, v would be the same for an $n = 15$ well program (provided that a success returned at least $5x$ or five times the well cost) because the program would still fail if the first 10 wells were dry holes. The program depends on a success somewhere in the first 10 wells in order to finance the remaining wells in the program.

Another case which equation (5.3) would apply to would be if the revenue stream is divorced from the budgeting. It may be the case that revenue generated by successes is not channeled directly back into the program. This would normally be the case when examining viability for a single operational office for a company divided into several operational divisions. The variability in total discoveries between divisions over a short period is huge because discoveries are infrequent and lognormally distributed. Hence, the budgeting back to the divisions is smoothed by corporate strategic planning above the local level. In this sense, equation (5.3) may be used rather than trying to incorporate return from successes as in the methods that follow.

Multiple Successes — Binomial Process

Sometimes the exploration firm is interested in analyzing the long-term viability of a program that at some point must begin to support itself if it is to remain perpetuated. For example, the firm may choose an exploration strategy that will take many years to complete and requires successes along the way. There may be enough initial capital for the first year or two, after which revenue from successes will be needed to continue. In this respect, it is possible that the program could fail even after an early success if it is followed by too many dry holes. Viability analysis then needs to extend into the long-term, past what can be determined from equation (5.3). The following approaches therefore incorporate the present worth return from successes into the capital that is available to perpetuate the exploration strategy.

Assuming that the average chance of test well success, p, remains constant during the life of the n well program, the program can be modeled as a binomial process. A binomial process consists of sequential trials, each with two possible outcomes of fixed probability. A classic example of a binomial process is counting the number of heads in a sequence of coin tosses. It is often useful to model a sequence of chance events as a binomial process so that associated probabilities can be easily calculated. An n well program can be thought of as a sequence of trials (test wells drilled), each with two possible outcomes (success or dry hole) of fixed probability (p and $1 - p$, respectively) so the probability of having m successes (where m can be any number between 0 and n) after all n wells have been drilled is

$$P(m|n) = \frac{n!}{m!(n-m)!} \, p^m(1 - p)^{n-m} \qquad (5.4)$$

where $P(m|n)$ is the probability of exactly m successes given that n test wells are drilled. Equation (5.4) is simply the probability function for the binomial distribution. If the net present worth return from a discovery is R then for any m such that

$$(C + mR)/x \leq n_m \leq [C + (m + 1)R]/x, \qquad (5.5)$$

where n_m is any number of wells satisfying the inequality, the probability that the ending capital is less than x (program was not viable) is

$$v = \sum_{i=0}^{m} P(i|n), \qquad (5.6)$$

which is the probability of having m or fewer successes during the life of the n well program. Note that m successes would generate mR revenue, which, added to the initial capital C, means that there is enough capital for $(C + mR)/x$ test wells. However, as shown in equation (5.5), this is less than the total number of wells n_m. Hence when n is defined as n_m in equation (5.5), there must be more than m successes if the ending capital is to be at least equal to the cost of the next test well, x. Note that if $n_m = [C/x]$ then $m = 0$ and the single term in equation (5.6) reduces to equation (5.3).

If there is enough capital for one more well at the end of the program, then technically the program remained viable. However, the exploration firm is more likely to be concerned with the probability of having at least some set ending capital representing the achievement of some profitability goal. If the goal was self-sufficiency then equation (5.6) may be rewritten

$$v = \sum_{i=0}^{m + [C/R]} P(i|n), \qquad (5.7)$$

where $[C/R]$ is the largest whole integer contained in C/R. Equation (5.7) gives the probability that the ending capital will be less than C, the initial capital, indicating that the program was not self-sufficient. Note that it is identical to equation (5.6) but with an additional $[C/R]$ discoveries needed to replace the initial capital. If some other minimum goal was set for ending capital, that figure can replace C in equation (5.7) to give the probability of not achieving the desired ending capital.

It is important to note that equations (5.4) through (5.7) calculate probabilities based on the assumption that the program is always completed (all n wells are drilled) regardless of their outcome. That is, the program is not abandoned even if the remaining capital plus return drops below x, the cost of the next test well. This approach then does not represent the true "ruin" case in which the program would have to be abandoned at any point where the capital available for the next well drops below x. These equations can be used for cases in which the capital available may become negative (that is, additional funding or loans are required) without abandoning the program.

Multiple Successes — True Ruin Case

The main limitation to equations (5.6) and (5.7) is that the program is not abandoned when there is no capital available for the next test well in the program. In the truly realistic case, the program may have to be abandoned because of a series of dry holes early in the program. For example, if the initial capital is large enough to drill 10 wells and the total long-term exploration program calls for 20 test wells over several years, then completion of the program will require at least one discovery (providing present worth return) in the first 10 wells. If all 10 were dry holes, the initial capital would be exhausted after the tenth well, and the program would have to be abandoned before any of the remaining 10 planned wells could be drilled.

In modeling the true ruin case, it is helpful to note that for a given set of C, x, R, and m, any value of n satisfying the inequality in equation (5.5) will yield the same viability measure or chance of ruin. For example, if there is capital enough for 10 wells and if the return from a success is more than the cost of another test well (which of course it should be) then a program that has 11 wells has the same viability measure as a 10 well program, because they both require a discovery in the first 10 test wells. In fact, if the return from a success is large enough to fund 10 more wells, then all 10 through 19 well programs have the same viability measure. However, a 20 well program with the same starting capital has a greater chance of ruin. Like the shorter programs, it must have a discovery in the first 10 test wells. but it must also have two discoveries by the program's end or the initial capital plus return will be exhausted. By a similar argument, all 20 through 29 well programs have the same viability or chance of ruin, but a 30 well program with the same starting capital has a slightly higher chance of ruin because there must be three discoveries by the program's end, the first two of which must occur before the tenth and twentieth wells.

This result seems to go against intuition since one might suppose that increasing the number of wells would increase the probability of total program success through the "law of averages." This would be true if the program would always be finished regardless of intermediate outcome (eqs. 5.6 and 5.7). But, as will be seen, when the beginning capital is held constant (as in reality it is), increasing the number of wells without sharing risk increases the probability of program failure.

In general, to remain truly viable and self-sufficient, the total capital available (which is remaining initial capital plus remaining present worth return from any discoveries) must always be greater than x, the cost of a test well. The m discoveries must fall in a manner that maintains this condition. There must be at least one discovery in the first $[C/x]$ test wells, at least two in the first $[(C + R)/x]$ test wells, at least three in the first $[(C + 2R)/x]$ test wells, and so on. The integer value of m can be determined from n, using equation (5.5).

If $m = 0$ then equation (5.3) gives the applicable viability measure, v_0 (a subscript will now be added to the viability measure, v, to indicate components 0 through m). If $m > 0$ then the program can become unviable only on some well numbered $[(C + iR)/x]$ where i could be any integer between 0 and m describing the number of successes. If there are enough wells so that $m = 1$, then it is possible to go broke even after a success, and the probability that the program will not be viable is

$$v_1 = v_0 + P(1 \mid n_0)P(0 \mid n_1 - n_0), \qquad (5.8)$$

which is derived from adding the probability that the program was not viable after n_0 wells, v_0, from equation (5.3), and the probability that the program was not viable but had exactly one success. This latter term is the probability of exactly one success in the first n_0 wells, $P(1 \mid n_0)$ times the probability of zero successes in the next $n_1 - n_0$ wells, $P(0 \mid n_1 - n_0)$ or $(1 - p)^{n_1 - n_0}$.

If $m = 2$ then the capital can only drop below x on well number n_0, n_1, or n_2 and

$$v_2 = v_1 + \{P(2 \mid n_0)P(0 \mid n_1 - n_0) +$$
$$P(1 \mid n_0)P(1 \mid n_1 - n_0)\} P(0 \mid n_2 - n_1). \qquad (5.9)$$

This equation reflects the possibilities of becoming unviable with zero or one success, v_1, plus the probability of becoming unviable with exactly two successes. The two successes can either be in the first n_0 wells or one in the first n_0 wells and one in the next $n_1 - n_0$ wells to remain viable through n_1 wells. If the last $n_2 - n_1$ wells have no successes, then the available capital would be less than x and ruin would result.

The viability measure or probability of ruin can be similarly constructed for $m = 3$ by enumerating all possible occurrences of three successes, allowing the program to remain viable through n_2 wells followed by $n_3 - n_2$ dry holes. However, as m increases, the calculations become much more complex as terms are enumerated and summed and probability intersections are subtracted. Furthermore, the probabilities for the subsequent terms tend to become more remote. Because of this, for $m > 2$, which is probably uncommon for reasonable programs designed for the foreseeable future, values of v can be accurately and conveniently estimated by the recursive formula:

$$v_m = v_{m-1} + (1 - v_{m-1})P(m \mid n_{m-1})P(0 \mid n_m - n_{m-1}). \qquad (5.10)$$

This adds the probability of ruin with fewer than m successes, v_{m-1}, to an estimate of the probability of ruin with exactly m successes. This latter term multiplies the probability of m successes in n_{m-1} wells, $P(m \mid n_{m-1})$, by an estimate of the probability that the m successes fell in a manner which kept the program viable through n_{m-1} wells, $(1 - v_{m-1})$, and the probability that the last $n_m - n_{m-1}$ wells are dry holes, $P(0 \mid n_m - n_{m-1})$.

The actual process of determining whether the capital available after each test well outcome has dropped below the cost of another test well is conceptually quite simple. A small simulation model can be programmed on a computer with very little effort. Figure 5.1 is a schematic for just such a model. In this diagram, rectangles represent calculations, diamonds represent conditional tests (transfers), and ovals indicate starting or stopping points. The cycle starts with drilling the first test well, and the available capital is reduced by the cost of the test well. Next the well must be tested to see if it was a success. This test can be done quite simply by drawing a uniform random number between zero and one within the computer and seeing if it is greater than or less than the specified decimal chance of success. If it is greater than the chance of success, the well was a dry hole, otherwise it was a discovery, and the available capital is increased by the present worth return of a discovery. In either case, the available capital is compared to the cost of the next text well. If insufficient capital is available, the program has failed to remain viable. If capital is sufficient for the next well, a test is performed to see if all wells in the program have been drilled. If they have, the program remained viable and the iteration ends. Otherwise, the next well in the program is drilled, and the cycle repeats. Iterations through the entire n well program are repeated several times in a "Monte Carlo" simulation, and the number of failed and successful programs is accumulated and printed out. The probability that the program will remain viable can then be estimated from the number of failed and successful programs.

A simple simulation model such as that shown in figure 5.1 would allow the removal of several limiting assumptions. Each test well to be drilled can have its own chance of success, cost, and return (either as constants or drawn from random

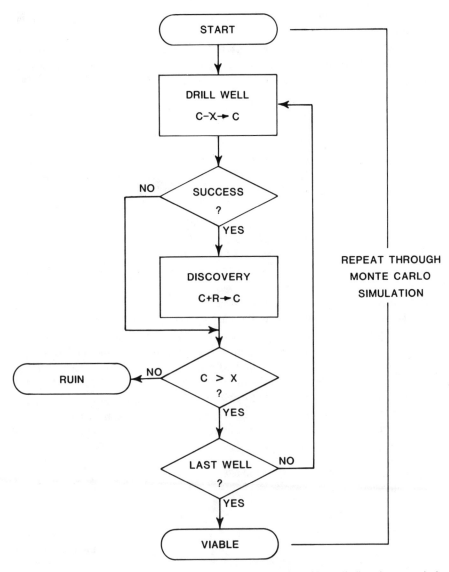

Figure 5.1. Flowsheet for a simple Monte Carlo simulation model, which will allow the removal of several limiting assumptions for calculating viability for a multiwell program.

distributions), and the simulation program can cycle through them in the order in which they are to be drilled.

Using equations (5.3) through (5.10) or a simulation model similar to figure 5.1, viability measures should be available for any multiwell program according to the applicable circumstances.

Example Viability Analysis

As an illustrative example, suppose the average test well cost is $300 thousand and there is initial capital of $3 million. If the expected present worth of a discovery is $3 million and the chance of test well success is 20% then:

$$
\begin{aligned}
x &= \$3,000,000 \\
C &= \$3,000,000 \\
R &= \$3,000,000 \\
p &= 0.20
\end{aligned}
$$

From equation (5.3) the probability of zero successes in $C/x = 10$ wells is 0.107.

Table 5.1 shows the results for various values of n from equation (5.6) in which all n wells are drilled regardless of intermediate outcomes and for the true ruin case (eqs. 5.3, 5.8–5.10) in which the program is abandoned if available capital falls below x. The fundamental difference in results is important. If the program is always completed even if external funding is needed as in case II of table 5.1, the probability of program failure decreases with additional wells, as might be assumed from the law of averages. However, in the more realistic case I, the program must be abandoned whenever the available capital is too small to continue,

Table 5.1   Example Viability Analysis

Number Wells (n)	Minimum Successes ($m + 1$)	Case I Prob. of ruin (%) (program is abandoned, eqs. 5.8–10)	Case II Prob. of ruin (%) (program is completed, eq. 5.6)
10	1	10.7	10.7
20	2	13.6	6.9
30	3	14.7	4.4
40	4	15.5	2.8
50	5	15.9	1.8

and the probability of program failure increases as the number of wells increases. This also follows because the program attempts to complete more and more wells with no additional viability resources.

It is important to remember that when performing any sort of analysis, the applicable circumstances for ending the program should be accurately represented. An important point to keep in mind when performing any form of viability analysis is that the applicable circumstances for ending the program should be accurately represented. Otherwise, the probability of program failure may be understated. Note that in this example the differences in stated probabilities are quite significant: in case I increasing from about 11% to about 16%, in case II decreasing from about 11% to less than 2%. If 2% had been considered an acceptably low probability of program failure, the analysis could have supported a wrong decision if the wrong failure criterion had been used.

Risk Sharing Techniques

The above foundations for evaluating the probability of program failure call to mind a strategy for sharing risk. If the probability of program failure is unacceptably high, a "dilution factor" can be determined with which the desired working interest can be found and the probability of program failure can be lowered. If the decision maker knows how low the maximum acceptable chance of program ruin should be, the maximum working interest allowable for maintaining this security can be easily found.

The variable x was used in the preceding section to indicate the cost of one test well. If x is the total test well cost then it needs to be multiplied by the working interest to find the drain on available capital. Hence equation (5.3) can be rewritten as (dropping the brackets for the moment)

$$v = (1 - p)^{C/ix} \qquad (5.11)$$

where i is the working interest expressed as a fraction. Rearranging equation (5.11) yields

$$i = \frac{C \log (1 - p)}{x \log v} . \qquad (5.12)$$

Now the desired working interest can be directly determined for a set value of v, the probability of program failure.

In a preceding example, C was 10, x was 1, and p was 0.15. If the working interest is 100% ($i = 1$) then equation (5.11) yields $v = 0.196$), nearly a 20% chance of program failure. Since this high probability of ruin is certain to be unacceptable, the explorationist must then decide what is acceptable. Suppose the probability of program ruin must be reduced to 5%, that is, $v = 0.05$. Substituting this value into equation (5.12) yields $i = 0.542$ or 54.2%. But now $C/ix = 18.45$ and because only a whole number of wells can be drilled, let $C/ix = 19$ and hence $i = 10/19$ or 52.63% and $v = 0.0456$. It has now been determined that 52.63% is the maximum working interest that would still allow for less than a 5% chance of program failure.

By repeating this approach for various values of v it is possible to construct a table of maximum working interest and number of wells versus the probability of program failure as shown in table 5.2. Note how quickly the probability of program failure is reduced by dilution. For this example it was possible to reduce the probability of program failure from about 20% to about 1% by participating with 34.5% interest in 29 wells instead of 100% interest in 10 wells.

Table 5.2 Maximum Allowable Working Interest for Example Program
($C = 10$, $x = 1$, $p = 1.5$)

Prob. of Program Failure (%)	Maximum W.I. (%)	Number of Wells (C/ix)
20	100.0	10
10	66.7	15
5	54.2	19
2	41.7	24
1	34.5	29
0.5	30.3	33

The same basic approach can be used for applying working interest dilution factors to the multiple success cases (eqs. 5.6 through 5.10). A new value of n is chosen (or else the same n wells as before are drilled but with less working interest), ix is substituted for x, this time in equation (5.5), and m is determined. The direct calculation of the dilution factor for a set risk level, v, is made mathematically complicated by the stepwise calculation of m and the combinatorial in equation (5.4). However, because the equations themselves are easily evaluated and because values of n must be discrete, an iterative search by a computer program will quickly converge to the value of n and associated working interest that yield the desired level of risk aversion. Similarly, an iterative search substituting ix for x in the simulation model in figure 5.1 will yield the desired working interest for risk aversion.

Viability as a Measure of Risk for Planning

The viability measure or chance of program failure can be controlled by strategic considerations. The effect of policy for determining when the program should be terminated (sometimes a one choice "decision") was demonstrated in table 5.1. Table 5.1 also showed that the number of wells affects the chance of program success. As was demonstrated in the preceding section, dilution through reducing working interest was a quantifiable method of increasing program viability. Similarly, dry hole and acreage contributions affect the measure of viability. If the test wells have different success ratios, cost, and returns, then the order in which the wells are drilled will have an effect on the viability measure. This effect can be determined by the simulation approach outlined in figure 5.1. Hence, it follows that the viability approach can be used not only to evaluate a program but also to choose a strategy that will provide the desired level of viability. Arps and Arps [1], Ramsey [2], and Greenwalt [3] give formulations for determining venture participation based on the probability of ruin.

The viability measure is useful not only as criteria for determining a particular exploration program within a play or basin but also as a risk measure for the program as input to the regional or corporate strategic planning level. At the strategic level, many programs must be combined into a portfolio of investment alternatives where the exchange between risk and expected return can be optimized with strategic goals in mind. In this way the corporate overall strategic plan can integrate programs of varying size and risk into a large exploration program that maximizes return to the firm while keeping cash flow problems at bay.

The Portfolio Approach

In the past, petroleum exploration has, out of necessity, been closely concerned with the physical environment. Effective search for deposits and efficient production will always require state of the art science and engineering. Yet increasing economic pressure from the business world of pricing, supply, costs, and interest rates will give added incentive to the development of comprehensive business planning for exploration programs. Corporations intent on growth or even survival will give added emphasis to business criteria, such as risk profiles, viability measures, cash flow timing, and portfolio efficiency when choosing geological plays and prospects. The traditional objectives of increasing the company reserve base at all costs or examining expected values, based only on long-term exploration programs and long-term economic stability, will need to be examined in the context of risk and financial feasibility. Portfolio analysis attempts to do this by aiding the decision maker in evaluating exchanges between return and risk by viewing the petroleum exploration industry as a portfolio of investment alternatives.

First, the exploration program must be organized into plays or some other local unit so that modular planning concepts can be applied. This will allow each play to be analyzed and the characteristics determined. Historical data can give an indication, in mature plays, of the distribution of field sizes to be found and the associated chance of test well success (as discussed in chapters 4 and 6). These data combined with information that is generated in the exploration process will provide the input for the portfolio approach.

A few simple matrices can be used as an initial method of analysis using the portfolio approach. The first matrix, shown in figure 5.2, compares risk to potential reserves. After the exploration program has been broken down by play, the average risk for each of the typical prospects can be used to calibrate the matrix for both the risk and potential field size scales. As each of the typical prospects are plotted, a pattern should develop that should give the trend of the program. This trend would generally run from the upper left to the lower right of the matrix. The position of the trend can be used to judge the risk potential of the program. The expected set of investments would probably appear in figure 5.2 in region II where the smaller field sizes are found with less risk. Region I represents less attractive alternatives, region III represents highly attractive investments.

A second matrix, shown in figure 5.3, compares total investment to return.

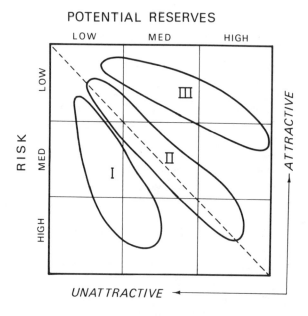

Figure 5.2. Matrix for risk versus potential reserves analysis for exploration programs.

SOURCE: Reprinted by permission of SPE-AIME from Quick and Buck 1983. Copyright 1983, SPE-AIME.

This matrix requires that the typical prospect or total play economic analysis be calculated. The range of results are then determined so that the matrix can be calibrated to fit the low, medium, and high categories. The points for each play are then plotted on the matrix and compared to the diagonal line that is calibrated to reflect the corporate acceptable rate of return. Similar to figure 5.2, region I, II, and III represent, respectively, unattractive, mixed average, and highly attractive investments.

Typically of course, investments with high expected return and high top end potential carry with them higher risk. Portfolio theory capitalizes on this basic law of business to locate the efficient frontier portfolio (so named for its position when graphed, see figure 5.4). The efficient frontier is the set of investments that give the maximum expected return for any given risk level. Conversely, it is also the set of alternatives giving the minimum risk for a given expected return. For example, if prospect A will yield an expected return of $1 million with 20% chance of success and prospect B will yield a return of $500 thousand for the same investment amount and same chance of success then prospect B cannot be along the efficient frontier because prospect A provides greater return at the same risk level. Prospect B cannot enter the efficient frontier unless there are no more prospects with 20% or greater chance of success with expected return given success greater than $500

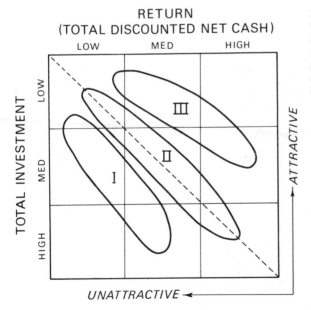

RETURN
(TOTAL DISCOUNTED NET CASH)

LOW   MED   HIGH

TOTAL INVESTMENT — LOW / MED / HIGH

III

II

I

ATTRACTIVE →

UNATTRACTIVE ◄——

Figure 5.3. Matrix compares total investment to return for exploration programs.

SOURCE: Reprinted by permission of SPE-AIME from Quick and Buck 1983. Copyright 1983, SPE-AIME.

thousand for the same investment (that is, after prospect A has been tested). Hence, any optimum exploration program must be taken from some point on the efficient frontier portfolio. The particular point (or points) chosen from the efficient frontier for the exploration strategy will depend on the desired risk level.

If potential return is plotted against risk for several prospects or plays then a portfolio emerges according to the grouping of alternatives as in the regions marked I and II in figure 5.4. Risk for multiwell programs can be based on the viability measures. The diagonal line on the graph represents the tendency of higher risk associated with higher returns. Portfolio I might represent a grouping of mature domestic prospects or programs while portfolio II might represent the higher risk and top end potential of international frontier plays. The efficient frontier can be drawn from inspecting along the edge that gives the highest return for a given risk level. In figure 5.4, the efficient frontiers for portfolios I and II are shown by the solid lines. Note that there are no more points in the portfolios giving higher return for any risk level than along the efficient frontier.

It may be that a single alternative in the portfolio cannot be repeated indefinitely. In fact, with petroleum exploration, alternatives are usually one-shot prospects or programs that, once completed or abandoned, will not be exactly repeated. A portfolio approach can still be used in this case. However, the efficient frontier will shift to include only the remaining alternatives as those at the extreme edge are exploited. Hence, all of the same concepts apply.

Locating the efficient frontier of exploration alternatives is fairly straightforward conceptually (although it can pose a tremendous evaluation problem in less

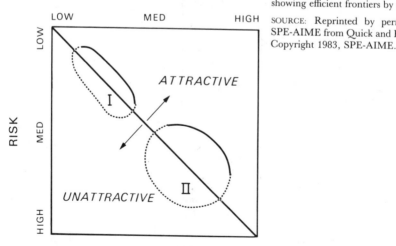

Figure 5.4. Investment portfolios showing efficient frontiers by solid lines.

SOURCE: Reprinted by permission of SPE-AIME from Quick and Buck 1983. Copyright 1983, SPE-AIME.

mature areas) but choosing where along the efficient frontier to be requires coordination of strategic goals. The traditional approach in portfolio analysis calls for the construction of a utility curve to represent the firm's aversion to risk. Grayson [4], Newendorp [5], and Harbaugh, Doveton, and Davis [6] give good discussions of the use of utility theory in oil and gas exploration.

There is strong argument for using utility curves for portfolio selection because they do provide correct theoretical basis. However, there are often large obstacles that must be overcome in deriving the actual utility function for a firm. Almost without exception, the utility curve, which was intended to add objectivity to decisions about risk, must be constructed in a subjective manner such as through indifference point interviews or through interpretation of past decisions. Also management participation is often minimal because of the complicated mathematical approach and because inconsistencies are often revealed over time (actually, the inconsistency should be expected since the company's or individual's aversion to risk is not constant over time). The problem then is not the subjectivity but the fact that it is made complex and is once removed from the decision maker.

A simpler but much less quantifiable method for handling risk is to include a desired risk profile directly into the strategy objectives. That is, since risk is a fact of life in petroleum exploration and because higher returns are associated with higher risk, the desired risk profile can be determined by the strategic planner's perception of the firm's current position for shouldering risk.

Typically, investment alternatives can be conveniently classified as in figure 5.5 which is similar to a three-dimensional version of figures 5.2 and 5.3. The cor-

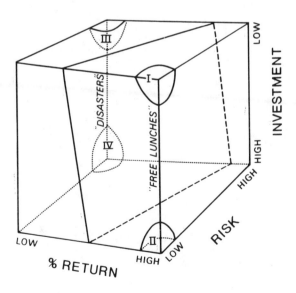

Figure 5.5. Three-dimensional version of figures 5.2 and 5.3 showing classification of exploration investment alternatives.

SOURCE: Reprinted by permission of SPE-AIME from Quick and Buck 1983. Copyright 1983, SPE-AIME.

ners marked I and II represent "free lunch" alternatives that simply do not exist in oil exploration. Corners III and IV represent, respectively, small and large disasters that are to be avoided. Some surface, such as that which is cutting through figure 5.5, would be the division between acceptable and unacceptable projects and is conceptually based on the risk profile. Figure 5.6 shows the matrix divided along the profile surface distinguishing between the acceptable and unacceptable projects. A computer graphics display of projects on a three-dimensional diagram such as these would be revealing since it offers several attributes for comparison.

The risk profile should indicate not only those alternatives that are acceptable, such as did the efficient frontier, but also indicate those that are preferred. Hence, the desired risk profile is a strategic objective. One method of representing the risk profile is shown in figure 5.7. This graph of cumulative expenditures versus risk for the projects making up the exploration program indicates how the strategy is balanced or tuned to risk. Risk may be the chance of test well success if evaluating an exploration program composed entirely of prospects or it may be the viability measures (chance of program success) of several multiwell programs so that plays can be combined for the highest level exploration strategy within the region or company.

Curves for some typical strategies are shown in figure 5.7. Curve 1 might indicate a uniform or risk balanced strategy in which the total expenditure for alter-

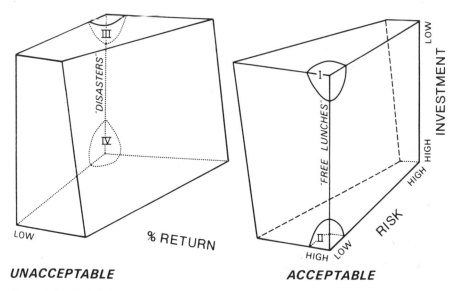

Figure 5.6.   Exploded view of figure 5.5 showing division between acceptable and unacceptable projects.

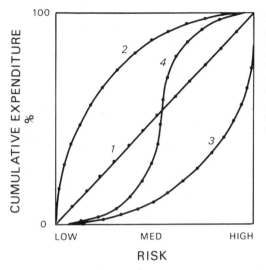

Figure 5.7. Risk profiles for various exploration programs showing typical strategies. Curve 1: uniform or risk balanced strategy, curve 2: conservative or risk adverse strategy, curve 3: aggressive or risk seeking strategy, curve 4: middle of the road or medium risk strategy.

SOURCE: Reprinted by permission of SPE-AIME from Quick and Buck 1983. Copyright 1983, SPE-AIME.

natives is spread evenly over all risk levels. Curve 2 might represent a conservative or risk averse strategy since the largest portion of expenditures is in low risk alternatives. An aggressive or risk seeking strategy would look like curve 3. A middle of the road strategy with most expenditures in medium risk projects would resemble curve 4.

The risk as shown in figure 5.7 gives an immediate graphical means of communicating the thrust of an exploration strategy. Through viability analysis and risk profile selection, the strategic planning can be broken down to any level from corporate to region to play to prospects for separate analysis and consolidation.

Portfolio grouping is a tool that can be used to bridge the gap betwen corporate planners and the operating entities. The operational exploration management can convey to corporate management the variety of strategies and options available to them. When the planners and top managers have chosen the strategic objectives in terms of programs and plays that should be followed, it is their responsibility to allocate the resources to carry out these programs. This often requires a long-term commitment since most programs operate in five- to ten-year periods.

Viability analysis provides a means of quantifying business risk at all levels as input to a portfolio approach for managing risk while setting exploration goals. Viewing exploration as a portfolio of investment opportunities and displaying the key attributes of exploration programs and plays will enable strategic planning at the corporate level to be integrated with the exploration strategies at the operational level.

References

1. Arps, J.J. and Arps, J.L.: "Prudent Risk-Taking," *J. Pet. Tech.* (July 1974) 26, no. 7, 711–716.
2. Ramsey, J.B.: "The Economics of Oil Exploration: A Probability-of-Ruin Approach," *Energy Economics* (Jan. 1980) 2, no. 1, 14–30.
3. Greenwalt, W.A.: "Determining Venture Participation," *J. Pet. Tech.* (Nov. 1981) 33, no. 11, 2189–2195.
4. Grayson, C.J., Jr.: *Decisions Under Uncertainty: Drilling Decisions by Oil and Gas Operators*, Harvard Business School, Cambridge, MA (1960).
5. Newendorp, P.D.: "Application of Utility Theory to Drilling Investment Decisions," D. Engr. thesis, University of Oklahoma, Norman, OK (1967).
6. Harbaugh, J.W., Doveton, J.H., and Davis, J.C.: *Probability Methods in Oil Exploration*, John Wiley & Sons, New York (1977).

6

Strategic Analysis: Strategy Modeling

Diversity of Exploration Approaches

The proliferation of different approaches to petroleum exploration used by oil companies should be evidence enough that there is no single best exploration strategy. Some companies try to maintain large acreage holdings, leasing all they can afford of domestic or frontier areas, while other companies are more selective about what they lease, buying only what they have immediate plans to test. Some companies place large stakes in geological and geophysical work, using the most advanced state of the art techniques or even developing new technology in the hopes of not only increasing their test well success ratio but also of making larger discoveries. Other companies seem to forsake all but the most traditional front-end geology and instead channel more money into test wells or acreage. Also, the variance in preferred working interest among companies indicates different attitudes to risk sharing. Companies also set different policies for farmouts, dry hole contributions, farmins and partnerships.

Among the most important decisions is the choice of plays. Companies may choose among plays ranging from expensive, high risk international areas to low cost, low risk domestic basins. The international areas quite often require not only huge front-end investment and years to payback but also involve considerable political risk. At the other extreme, many domestic basins are so extensively explored that large discoveries are very unlikely, and explorationists must depend on low risk, but smaller, "bread and butter" discoveries. Between these extremes is the full spectrum available to the explorationist. Even in well-known and lucrative

Portions of Chapter 6 originally appeared in altered form in *AAPG Bulletin*; copyright 1982, AAPG [Jones, D.A., Buck, N.A., and Kelsey, J.H.: "Model to Evaluate Exploration Strategies," *AAPG Bull.* (March 1982) 66, no. 3.] and later in *World Oil*.

areas, such as the Permian basin, one can find diverse approaches, with some companies sticking close to existing production, others stepping out into rank wildcat areas; some companies opening up new provinces with deep well tests, others finishing out shallow plays originally discovered 70 years earlier.

Even the mixes of plays and prospects within single companies show much variety. Each company tries to maintain a careful balance between tried and true areas, in which long years of investments were required to establish a position, and frontier areas that require exploration now in order to have a profitable position tomorrow. Acreage bonuses have a tendency to skyrocket overnight with the discovery of a significant field and seldom come down again if there is oil in the area. Also, as has been shown numerous times, the largest fields tend to be discovered very early in a basin's exploration history.

Obviously, the financial environment in which a company must operate greatly affects the course of its exploration program. During highly favorable periods, such as 1973–75 and 1979–81, brought on by rapid price increases, the industry enjoyed highly aggressive exploration both domestically and worldwide, in established as well as in new provinces. In industry depressed periods, such as 1982–83, companies had to rethink their riskier concerns, pulling out of areas no longer deemed profitable.

A company also has an individual financial environment that closely affects long- and short-term planning and is affected by the success or failure of past exploration as well as other company concerns. Therefore, even if all external influences remained constant, which indeed they do not, a company's outlook, and hence its strategy, can still change drastically. A small independent company reaping peak revenue from its share in a major discovery may be a more aggressive risk-taker than a much larger company severely limited by cash flow problems. Hence, it is no surprise that even a single company will, over a number of years, employ diverse exploration approaches, partially due to external trends and influences and partially due to shifting internal culture and constraints.

The Need for Strategic Modeling

The complexity of the technical, economic, and corporate environment in which exploration operates dictates that the choice of a strategy is too involved to be decided merely by tradition or momentum of existing approaches or simple rule of thumb policies. The choice is too important to be left to intuition or faith. Instead, the explorationist must draw on his understanding of the environment gleaned from experience, current information, and new ideas. It is these last two components that strategic modeling supplements. Information must be current and not mere historical data. That is, information must be related to exploration so that it helps illuminate strategic choice. For example, reliable forecasts of future oil discoveries within a basin are of more use in decision making than compiled statistics of historical discoveries. Similarly, a single figure of total undiscovered oil in a basin, while of use to government energy policy setting or macroeconomics, is of less use to explorationists than when given in the form of average discovery size by play or depth and related to the time series of exploration effort required. Also new ideas are only valuable to the explorationist once evaluated and made practicable. Therefore, if a model of the exploration environment can be incorporated into a theoretical basis for evaluating exploration alternatives, a means for utilizing current information and exercising new ideas is made tangible.

The extreme variance of outcomes experienced in oil exploration decreases the usefulness of expected value or average case evaluations. It has been estimated that the probability of a wildcat well drilled in active areas in the continental United States discovering a profitable quantity of hydrocarbons is typically no greater than one in nine. Considering that the highest success ratios, though probably not the largest remaining deposits, for American oil companies are achieved on shore in the continental United States, it is evident that even a multiple well program still must meet with great risk. This risk is even more pronounced in important frontier plays where the expected value of a project's rate of return (which would usually be quite attractive) may be of less use in decision making than the probability that the project will achieve some minimally acceptable rate of return. The latter may be too small, even with a multiple well program, for any but the largest oil companies.

Even when a test well discovers a petroleum deposit (an event which may be considered historically uncommon) a great deal of outcome variance still exists for, as has been often demonstrated, petroleum discoveries are highly skewed and may be modeled as lognormal. Even within a single play, the larger fields may yield a thousand times as much oil as the smaller fields that are found a hundred times more often. The two compounding sources of variance — the chance of success and the distribution of discovery size — give added impetus to the development of strategic modeling techniques.

The great complexity of the exploration environment and the extreme variance of program outcome variability, which makes even hindsight evaluation difficult, seem to combine to encourage two prevalent attitudes in petroleum exploration. The first approach, which may be called "trial and error," must to some extent always exist in order to open up new provinces. Trial and error has been significantly reduced by each development of modern geology. However, these advances are constantly offset by the continual exhaustion of known provinces. Hence, trial and error becomes decreasingly feasible, especially in light of increasing drilling costs due to deeper discovery depths. If it is possible to accurately evaluate the economics of an exploration program through strategic modeling, that is, pretest it, the reliance on actual trial and error drilling can possibly be reduced.

The second prevalent exploration attitude may be called the "sheep" or "gold rush" approach. It springs up quite naturally as oil companies quickly move into areas where other companies have experienced success. Obviously, it is lucrative to move into such areas as soon as possible. However, a few things happen that decrease the comparative effectiveness of relying too heavily on the sheep approach. As has been shown, the largest fields are typically found very early in a basin's history. Indeed they quite often are the first significant discoveries and the ones that most inspire activity by other companies. The ensuing flurry of activity quickly increases the acreage bonus prices to several times what they were a few test wells earlier. Hence, companies relying totally on following the latest discoveries find their effectiveness reduced by paying higher bonuses for acreage holding smaller discoveries. The objectivity added by strategic modeling will help indicate the profitability of reacting to industry trends and the resulting gold rush effects.

The main objective of strategic modeling is to provide support for decision making by the explorationist or exploration management. Figure 6.1 demonstrates the role of strategic modeling in supporting the exploration manager. It is the responsibility of exploration management to arrive at the choice of an exploration strategy. Management relies on the experience and intuition of company personnel to recommend strategies based on their obervations of the exploration environment. The three boxes on the right of the diagram represent the main steps to strategic modeling and demonstrate how the choice of an exploration strategy can be supported. The steps certainly are not alien to exploration management; the briefest prospect evaluations go through exactly the same basic steps as complete strategic modeling. Environmental analysis makes use of observable attributes to derive parameters describing the exploration environment. From this description, controllable variables can be identified and a means of modeling these attributes is synthesized. Strategic alternatives are generated and tested (evaluated) with the model. The resulting deductions are then used to support the choice of an exploration strategy. The diagram in figure 6.1 has been generalized so that it represents equally any level of detail and aggregation.

The greatest potential for decision support lies with pretesting exploration alternatives objectively, while avoiding the infeasibility of testing exploration strategies by trial and error drilling. Strategic modeling attempts to do this by combining environmental analysis with a conceptual model of exploration alternatives available to the decision maker.

Forecasting Future Discovery Rates

The most important environmental attribute of a prospective basin or play is the distribution of future discoveries. This is also the most difficult to forecast. But discovery forecasting is absolutely necessary to environmental modeling. Also past

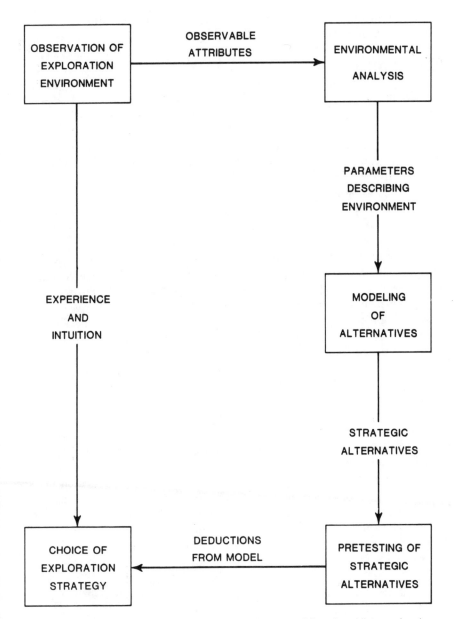

Figure 6.1. Flowsheet showing the role of strategic modeling in aiding exploration management's decision methods.

studies indicate that it is possible to successfully forecast future discoveries primarily because, despite economic and political influences, the discovery of petroleum deposits is essentially a physical process. It is usually safe to assume that if a hole is drilled to a certain depth, then any hydrocarbon deposits of economic size, which are positioned at the target depth or shallower, will be discovered. Also, whether there is a discovery or not, there should be no need to drill a hole in the same place except to a deeper depth. Hence, some portion of the basin is exhausted every time a hole is made in the ground. Admittedly, there are some exceptions. The log analysis may have been misinterpreted or it may not have been done for depths shallower than the objective layer and reserves may have been overlooked. But generally this simple-minded assumption is strong enough to provide a very useful basis for forecasting future discoveries based on well activity.

For clarity, a special note should be made here that only those forecasts expressed as a function of effort are being discussed. Forecasts of actual effort or industry activity are heavily dependent on socioeconomic influences such as political policy, changes in drilling technology, supply-demand, etc., and hence are more open to interpretation and opinion. The influences need not necessarily concern the explorationist because, as will be seen for the normal planning horizon for an individual exploration firm, such forecasts as total U.S. production per year are not as useful as forecasts of average reserves per test well drilled for the next 100 test wells drilled in a particular basin. Studies concerned with forecasting industry activity or ultimate reserves are of use for long-term planning or government policy setting but not for regional strategic modeling. Explorationists need to know what they can expect to find if embarking on a particular exploration program within a certain basin or play.

The objectives here are threefold: (1) to survey some forecasting studies that used a variety of methods to address different resource appraisal questions, (2) to aid explorationists in interpreting present and future studies by providing some background information about forecasting techniques and related assumptions, and (3) to provide a starting point and references for those who would like to undertake producing their own forecasts. Obviously, forecasts are most useful to explorationists when they can be applied to the actual decision process. Hence, those methods that can be used within the context of strategic analysis are emphasized rather than, for example, government studies aimed at assessing long-term world resources.

Graphical and Curve Fitting Techniques

The simplest and most common type of forecasting methods are empirical or graphical in nature. That is, they rely on statistical summation or graphical means to portray trends without strictly modeling the actual exploration process. Because of the physical nature of exploration (where variance is largely due to the natural distribution of oil and the level of drilling effort as opposed to socioeconomic systems) empirical methods based on level of effort can still give useful results. In fact, the largest portion of the variance of the resulting estimates for a particular level of effort is usually due to the lognormal nature of petroleum deposit sizes rather than socioeconomic influences.

Empirical methods (those which do not make geological or physical process connections) have the advantage of being, in relative terms, easy to undertake and understand. They assume that some mathematical function (or graphical curve) can be found to represent the relationship between reserves discovered and exploration effort. Exploration effort is usually measured in number of wells or footage from wells rather than time because activity, particularly in a specific area, may vary significantly from year to year. The curve is fitted by attempting to reduce the differences between points along the curve and corresponding data points observed from historical data. Once the curve has been fitted as well as possible to the observed data, it is then extrapolated over future effort. Figure 6.2 illustrates the basic form of a curve fitting model. The available history makes up the observable data. Discoveries are shown in the figure as cumulative in their simplest form. However, they could have been handled in a variety of methods, such as logarithm of cumulative discoveries, average discovery size for n well increments, average discovery size smoothed by moving averages (nearest n wells) or any other form that the forecaster felt was relevant or useful. The forecast function, usually taken from some known family of mathematical curves, is statistically fit to minimize the error components between data points and model values (see figure 6.2). Generally, the sum of the squares of these error components is minimized, known as a least squares fit.

Perhaps the most common empirical fit is linear regression in which the forecast function is a straight line fit by least squares. Often it is possible to transform the variables (such as by taking the logarithm of discoveries) in such a way that the relationship appears linear. This is often done because the method of least squares fit for a straight line function is relatively simple and straightforward to calculate and explain.

As an example of a simple linear regression curve fit forecast, figure 6.3 shows average discovery size fit to number of test wells drilled for the Big Horn basin in northern Wyoming. The basin's history was divided into two-year periods. The discovery sizes were averaged for each period and resulting averages were plotted against the number of wells drilled. Years are also indicated for reference purposes.

From a statistical standpoint, it may have been preferable to use periods, for example, of 100 test wells rather than years, since the number of test wells per year varies over time. However, the data source used here gave the discovery years and

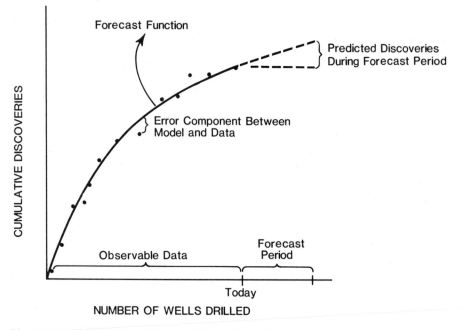

Figure 6.2. Empirical curve fit forecast of future discoveries.

the number of test wells drilled per year but not the actual data stream of test wells with matched discoveries. This illustrates an important point when forecasting. Often the forecaster must decide if more complete data will improve the forecasts enough to justify the additional time and cost of securing and verifying the additional data.

Note that the average discovery sizes are shown on a logarithmic scale in figure 6.3. A straight declining line on this graph would indicate an exponential decline in average field size. In 1935, after about 100 test wells had been drilled, the average discovery was around 10 million barrels of oil equivalent (BOE). By 1975, after about 1000 test wells had been drilled, the average discovery dropped to below half a million BOE. This decline should be expected because larger fields also have larger areas when viewed from the surface and were easily found with a few test wells.

While it is obvious that there has been an overall decline in average discovery size, an interesting thing happened in the early 1960s. The average discovery size appears to have increased over the next several hundred test wells. In fact, there was an increase due to the discovery of large stratigraphic traps in the Cottonwood Creek area of the basin. Hence, the entire pattern of large early discoveries and ex-

BIG HORN BASIN

DISCOVERY SIZE vs. WILDCATS DRILLED

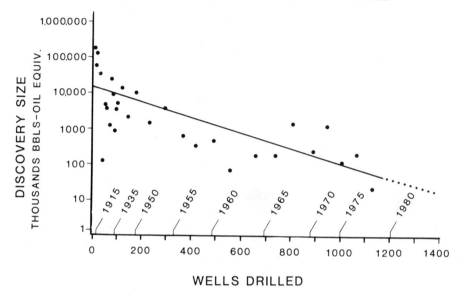

Figure 6.3.   Discovery size versus wildcats drilled, straight line curve fit of average discovery size versus number of wells drilled in Big Horn basin.

ponential decline begins again, as is the case when new plays or areas within the basin open up due to new geological information and drilling experience. Implied is that the forecaster should become as familiar as time allows with the history of the study areas in order to find assignable reasons why a particular technique would or would not work well in the study area.

Forcing a straight line fit is not advisable if some other function does a better job of representing the relationship. The curve fitting for most simple functions such as polynomials or hyperbolics can usually be worked out on a digital computer, and the arithmetic should not be a restriction if the functions truly seem to represent the underlying relationships. Whether an empirical fit will provide useful information at all is a more important consideration. The underlying exploration process must follow fairly straightforward development in the area and have sufficient maturity. Also, of considerable importance is whether reliance on the forecasts will be heavy enough to merit the additional research of more sophisticated methods.

Empirical and curve fitting methods appear to be most successful for high levels of aggregation, such as when looking at all of the United States or whole large basins where sudden shifts in industry activity and technology are smoothed over time. The advantages are ease of implementation, hence short development time, and ease of understanding and communication. The main disadvantages are the inability to handle sharp shifts in technology or activity and the dependence on relative maturity. Both of these limitations tend to rule out local studies of new prospective areas, which are of the most interest to explorationists.

Perhaps the most famous forecasts produced using empirical methods combined with expert judgment were those by Hubbert [1, 2]. Hubbert's main concern was with projecting yearly U.S. production, but in so doing, forecasts of discoveries were needed. Later studies by Wiorkowski [3] and Mayer, Silverman, Zeger, and Bruce [4] were also aimed at production projections and showed results reasonably close to Hubbert's for the new discoveries component. Schuenemeyer [5] in his comment to Wiorkowski's review of resource estimating techniques [6], gives a graphical comparison of estimates of U.S. ultimate oil production based on discovery data for several major curve fitting studies. Interesting to note is that most of them are consistent with Hubbert's 1962 projections. The reason is probably the high level of aggregation (total United States excluding offshore and Alaska), exploration maturity, and the physical nature of the exploration process. While studies with such high levels of aggregation are certainly useful for setting government energy policy and for long-range industry planning, they are of limited use to the operational decision making of an individual exploration firm. With this consideration in mind some techniques which are more useful in localized or immature areas will be discussed.

Exploration Process Models

Exploration process models include physical assumptions derived from the actual discovery process as part of their formulation. These models are designed for use in a single play or division of plays, such as in a basin or subbasin. They can be used to give detailed estimates either of ultimate discoveries or time series discoveries versus exploration effort. By modeling actual governing processes (such as spatial relationships of field areas and percent basin exhaustion) and technological considerations (such as industry exploration efficiency), the forecasts are given additional credibility over empirical or "black box" techniques. Though process model studies generally require more complete data and greater effort than simpler forecasting techniques, they show great promise for use by the individual explorationist or exploration firm.

Kaufman, Balcer, and Kruyt [7] present a process model built on probability theory for sampling from geologically homogeneous prospects as would be found in a single play. Their model makes use of the fact that large fields tend to be found early in the exploration process within the play. They assume that nature has deposited a discrete number of hydrocarbon fields with volumes drawn from a random probability distribution. Previous work by Kaufman [8] indicated that this distribution, which is highly skewed (there is a large proportion of very small fields and a small proportion of very large fields) could be closely approximated by the lognormal distribution. Hence, the exploration and discovery process could be equated to sampling relative to remaining fields proportional to size from a discrete number of fields to handle the decline in expected discovery size.

With sampling proportional to size, the probability, $P(B)$, that the next discovery will be of some given class size, B, is

$$P(B) = (KB)/S, \qquad (6.1)$$

where K is the number of undiscovered deposits of area B and S is the sum of areas of all undiscovered deposits. It is important to note that sampling is proportional to the sizes of fields discovered so far, because the largest fields tend to be discovered first. Building on this idea, the probability of success of a test well, p, was modeled as

$$p = kS/(kS + U), \qquad (6.2)$$

where U is the sum of the areas (alternatively, volume could be used for B, S, and U) of all undrilled prospects that are potentially identifiable by current exploration technology and k is an index to the difficulty of discovering deposits within the play. This index is a way of measuring efficiency, which is generally determined by the amount of geological information available for the play. The authors of this model go on to formulate physical possibilities for the arrival of new plays in which the above model would be a submodel for one play. This expansion, though beyond the scope of this survey, does much to improve the model's applicability to less mature basins or provinces where significant new plays are certain to be discovered.

The probability that some given sequence of discoveries will be made can be calculated as the products of the individual probabilities from equation (6.1). Using probability methods, Barouch and Kaufman [9] derived a likelihood function (probability density of sequences) for the parameters of the size distribution and number of deposits. Their expected value forecasts were then taken from the parameter set giving maximum likelihood.

Arps and Roberts [10] proposed an early model for the exploration process in which they equated the total oil in class size a fields discovered after w wells have been drilled, $F_a(w)$, to the ultimate oil, $F_a(\infty)$, multiplied by a decline factor based on maturity:

$$F_a(w) = F_a(\infty) [1 - \exp(- CwA/B)], \qquad (6.3)$$

where A is the area for a given field class size a and B is the total basin area. Schuenemeyer and Root [11] give a method for estimating the exploration efficiency, C, from past data, providing the only unknown in equation (6.3), which can then be used to forecast future reserves.

Drew, Schuenemeyer, and Root [12] developed a similar model based on geometrically derived basin exhaustion, recognizing that each test well drilled contributes to the total exhaustion of the basin whether a discovery is made or not. They modeled the fraction of deposits that have been found, f, within a play as

$$f = 1 - (1 - E/B)^C, \qquad (6.4)$$

where E is the total area exhausted, B is the total basin (or play) area, and C is exploration efficiency. If the forecaster is willing to make some assumptions about the shape and orientation of fields in the basin or play, the area exhausted by each well can be derived geometrically. For example, if fields are assumed to be circular in shape (a shape that works surprisingly well for many basins), the area exhausted by each well is equal to the area of the circular field for a given class size. Figure 6.4A illustrates in two dimensions that if the test well was located within the radius of the field, it would have been discovered by the well, so the area exhausted can be taken on the average as the field area. Note, however, that the area exhausted by the test well is not an exact physical property but is dependent on the field size being considered as shown in figure 6.4B where a field of smaller size may not have been discovered. This explains the dependence of field size in equations (6.3) and (6.4).

If fields are assumed to be elliptical in shape and randomly oriented, then a test well can still be within the outer radius of the ellipse without discovering the field (see figure 6.4C) depending on how the field is oriented. The exhaustion for a well can then be expressed by probability contours. If the well is within a distance equal to the inner radius of the ellipse, the probability of discovery is 100%. If the test well is outside the outer radius, there can be no discovery. At any distance in between, there is some probability less than 100% that the field is discovered depending on the distribution of orientation. Singer and Drew [13] showed how to calculate exhaustion for these elliptical targets, and Schuenemeyer and Drew [14] wrote a computer program to plot the exhaustion sequence.

If exhaustion is calculated for various class sizes and plotted against the number of wells drilled, then the fact that large fields are found early in the exploration process is clearly illustrated. Figure 6.5 shows what this plot would typically

A

Figure 6.4.   Area exhausted by test well for various field shapes. **A** indicates if test is located in the radius of the field; **B** indicates where a field of smaller size may not have been discovered; and **C** indicates random by oriented elliptical shape field not discovered.

B

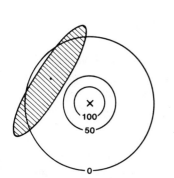

C

look like. Note that a basin that is 90% exhausted for large fields may be only 10% exhausted for small fields.

The development of basin exhaustion calculations and plotting methodology was an important by-product of these later models. Exhaustion mapping is of use to the explorationist even without the corresponding reserve forecasts for it allows summation of the current state of exploration for an entire play in an immediate and accessible format.

Volumetric/Geologic Analog Models

The curve fitting techniques and exploration process models discussed so far depend on information gained only through past exploration experience in the study

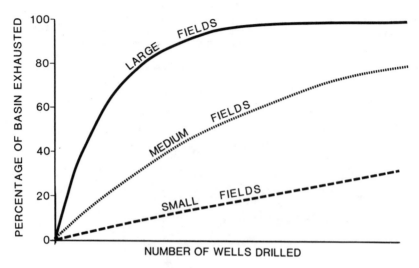

Figure 6.5.   Percent exhaustion versus wells drilled for three field size classifications.

SOURCE: Reprinted by permission of AAPG from Jones, Buck, and Kelsey 1982.

area. But if large fields are found early in the exploration of new areas then realistic estimates of reserves in unexplored areas are of great value to the explorationist. Obviously, all methods that bravely face estimation in unexplored areas must also face great uncertainty. Forecasting along frontier provinces is a topic that lends itself much easier to criticism than to development of techniques. But the explorationist must have some prospective estimation, even if dubious, if he is to move with industry instead of behind it.

The most common forecast techniques used in unexplored areas can be broadly termed volumetric geologic analog models because they make calculations scaled by basin volumes of sedimentary or source rock and are drawn from analogs to geologically similar explored basins. Generally, such studies require the following steps:

1. *Many basins of diverse geology but with some exploration history are categorized by geologic attributes.* Basins consisting of several distinct zones are first divided into zones or subbasins. This may be done subjectively by experienced geologists or by some statistical method such as factor analysis (simply stated, factor analysis derives a mathematical function for combining measurable variables, such as permeability, depth, age, etc., into a unitless "factor" that should yield similar values for similar basins or zones). Because all basins have some unique characteristics, a trade-off must be made between having too many categories, which impedes categorizing the unexplored study area, and too few categories,

which causes gross generalization and loss of distinguishing characteristics.

2. *The hydrocarbon content per unit volume for each category is calculated.* Note that this requires an estimate of total oil in place including past production, unrecoverable reserves, new or expanded pools, and new field discoveries for the analog basins. The uncertainty in later steps is large enough that the forecasts are not likely to be oversensitive to some variability to reserve estimates for these explored basins. Stated differently, as ambitious as this step sounds, the real inaccuracies arise in calculating content factors for the unexplored study area. Also, the forecaster can draw upon the vast amount of work done by the U.S.G.S. and the petroleum industry in estimating total recoverable reserves for partially explored U.S. basins.

3. *The unexplored study area is divided into basins or zones.* Each is then categorized according to similarities between what can be observed for the unexplored zones and characteristics of the mature zones.

4. *A model is applied to each unexplored zone.* This model relates the observed geological attributes (such as type of source rock, occurrence of traps, migration, etc.) to the recoverable resources per unit volume based on the mature zones and multiplies them by the volume of that zone.

This procedure, which may sound straightforward enough, always requires a great deal of expert judgment simply because of the great variability in reserves in place between two geologically similar basins. This is not hard to understand because the presence or absence of a single large field, which is determined by several simultaneous quirks of nature, greatly affects the observed measure of reserves per unit volume for an entire zone. Also, this method was primarily developed for calculating total recoverable reserves in the unexplored zones, a figure of less use to the explorationist than the probability and size distribution of discoveries in the new area. Hence, the explorationist must add another analog relating discovery size to that of a mature area.

Hendricks [15] described a geologic analog method similar to that given above to estimate hydrocarbon resources in inadequately explored areas. Jones [16], recognizing the problems with traditional volumetric/geologic analog models, attempted to increase the geologic input by incorporating the dynamics of oil generation, migration, and accumulation. His main emphasis was on trying to isolate the characteristics of highly prolific basin segments in spite of their geologic similarity to less productive basins or areas.

Jones compared highly productive segments in such geographically separate areas as southwest Iran and east Texas and identified the following four prerequisites for large oil accumulation:

1. Large traps
2. Large volume of effective reservoir rock in traps

3. Adequate source of hydrocarbons
4. Effective migration system

Next, he developed a descriptive model to quantify these four main attributes and combine them into an expression for reserves per cubic mile. His model was

$$\text{Reserves/cu mi} = a(R)\,(T)\,(P) \qquad (6.5)$$

where

$a = 26 \times 10^6$ (conversion for barrels to cubic miles)
R = fraction of basin that could contain producible reserves
T = fraction of basin that is in trap position
P = fraction of trap capacity that contains petroleum

R is calculated as the volume of capped reservoir (fraction of basin) times the recovery factor (barrels recovered per unit volume), T is calculated from trap volume divided by basin volume, and P is petroleum in traps divided by trap capacity.

This predictive model was then calibrated based on data from mature basins in which all of the components in equation (6.5) could be independently calculated. When applied to unexplored regions, the division into components helps maximize geologic input, but the forecaster is still left with the subjective estimation of R, the fraction of the basin that could contain recoverable reserves, and must fall back on expert judgment.

Econometric Models

The forecasting methods discussed so far were developed to relate the size of discoveries to exploration effort with the express objective of extrapolating into the future. This was based on the preceding assumption that exploration, when related to drilling effort, is largely a physical process. While this assumption is generally strong enough for the purposes of strategic modeling within the individual exploration firm, there are many agencies, particularly regulatory in nature, that cannot afford to make this assumption. Furthermore, they must go a step further and forecast future exploration effort and the effect of policy on encouraging (or discouraging) future effort. Because the methods developed for these purposes are of potential use to the exploration firm, especially for long-term planning, they are briefly discussed here.

When viewed from the total industry perspective rather than the individual firm, macroeconomic considerations do much to determine the exploration effort level that is attractive to industry. Some influences can be controlled by national policy while others are virtually independent from U.S. concerns. More often, economic variables are controllable in part but are too complex and interrelated to be controlled (or sometimes even understood) wholly. Obviously, pricing, tax structure, market demand, and current production each have a large impact on the attractiveness of exploration. But separating the effect, even short term, of any of them alone is impossible because all are related, not only to each other but to other economic variables, such as availability of investment capital, inflation, political risk, etc. Carried a level deeper in complexity, economic and technological states can even affect discovery size versus level of effort. For example, if the price of oil rose high enough to justify state-of-the-art enhanced recovery methods (or if such methods became more economical) then the recoverable size of new and old discoveries as well as production rates would increase substantially. Theoretically, this new surge of petroleum on the market could then lower the price, perhaps to the point of making enhanced recovery less attractive and once again lowering recoverable discovery sizes.

Models that attempt to tie economic variables together, often recursively, according to observed or intuitive multivariate relationships described by mathematical functions draw upon the field of econometrics to estimate their parameters. Econometric models may be as theoretically simple as the least squares linear fit discussed previously or intricate enough to incorporate thousands of computer simulated variables and relationships. The latter, of course, are beyond the means of any but the largest oil companies or government regulatory agencies. However, some smaller, straightforward econometric models are potentially useful to exploration firms.

An early econometric discovery forecasting model proposed by Fisher [17] related the average discovery size for a given time period to the discovery size, success ratio, and price from the previous time period. Such models are said to be recursive. Fisher's model estimated the average field size for oil found in time period t as:

$$S_t = \frac{a_0 \, S_{t-1}^{a_1} \, F_{t-1}^{a_2}}{N_{t-1}^{a_3} \, P_t^{a_4}} \tag{6.6}$$

where F_{t-1} is the success ratio from the previous period, N_{t-1} is the average size of gas fields found in the previous period, and P_t is the price of oil in period t. The a_i values are parameters that are fit (in this case restricted to be positive) by econometric methods in such a way that the model represents past history. Fisher's work was extended by Erickson and Spann [18]. MacAvoy and Pindyck [19] gave another econometric model estimating discovery volumes based on geologic zones and the number of cumulative test wells drilled.

Summary of Forecasting Techniques

In this short survey of reserves potential forecasting, four general approaches or types of models were discussed. Table 6.1 gives a summary comparison of the four methods as they relate to strategic modeling.

Table 6.1   Comparison of Forecasting Approaches for Use in Strategic Modeling

	Required Level of Aggregation	Applicability to Frontier Areas	Physical Realism	Ease of Use	Detail of Resulting Forecasts
Curve fitting	large	low	low	high	low
Process models	small	low to medium	high	medium	high
Volumetric	small to medium	high	medium	medium	low
Econometric	medium to large	low	medium (good econ.)	low	medium

The simplest method was empirical in nature, relating some mathematical curve to an historical time series. It had the advantage of being easily employed and understood but required relative maturity and stability of the exploration history. Most disabling for strategic modeling purposes, empirical methods require high levels of aggregation and, hence, cannot yield detailed forecasts by play. Also discussed were discovery process model techniques that attempt to represent the actual physical exhaustion process. These techniques generally prove to be the most valuable for strategic modeling because they often provide sound, detailed forecasts of discovery sizes by play. The main disadvantage of the process models is that they require partial maturity of the basin or play because of their reliance on historical data. The only methods that were applicable to frontier areas were the volumetric models which attempt to forecast potential in unexplored areas by analog comparison to explored basins. As must be expected for methods applied to frontier provinces, the volumetric methods give highly variable results and do not yield detailed forecasts. Econometric methods were mentioned since they attempt to include socioeconomic influences on industry effort. But as was mentioned, these do not play an important part in short-term forecasting for exploration firms who only need to forecast discovery size given any level of effort.

Simulating Exploration Strategies

Representing the Exploration Environment

Pretesting of exploration strategies requires modeling of the physical, technological, and economic environment in which the strategy must perform. The relative value of available exploration strategies is determined by such environmental attributes as local reserve values, operational costs, available company resources, government restrictions, local contract specifications, and quality of geological information as well as the more obvious influence of physical geology and expected reserves discoveries. All of these things actively affect the trade-off between maximizing company reserve additions and minimizing finding cost per barrel. As was shown in figure 6.1, strategic modeling, regardless of detail, begins with analysis of the environment based on observable phenomena. While the most influential single attribute may be the forecasting of discovery sizes, several other environmental variables must be considered if strategies are to be pretested by application to a simulated exploration environment.

The steps to the scientific process by which prospects are generated within an individual firm can provide insight to artificially representing the exploration environment. Quite often the process begins with reconnaissance scale geological or geophysical research where the emphasis is on locating areas that merit closer examination. These areas of interest are then searched for structural anomalies or other favorable conditions for trapping hydrocarbons. These local anomalies, or prospects, then are generally closely examined to estimate expected discovery volume and to locate potential test well sites. All this science should have two effects on the test well outcome: the success ratio should improve and the average discovery size should increase because the firm is better able to delineate trap volume and the larger prospects within the play will tend to be selected for testing.

With the current technology, however, it is safe to conclude that the effect of science must have some economic limit. For example, seismic lines can be shot on a smaller and smaller grid with the gain of more clearly outlining the structural anomaly but at the cost of increasing the geophysical survey cost, which in theory could approach the actual cost of drilling the test well. But at some point, only drilling the test well will really improve knowledge of the potential field. In fact, from a theoretical standpoint, it would be plausible to think of the test well as the next step in the geological analysis. For the moment, however, if the effects of geological and geophysical effort alone on success ratio are considered, then a curve of eventual diminishing returns can be hypothesized as shown in figure 6.6.

Note from the curve in figure 6.6 that there is some probability of test well success even without any preliminary analysis, that is, random drilling would still have some chance of discovering oil. This is evidenced by the many cases of serendipity

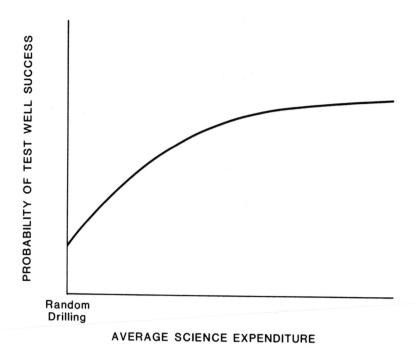

Figure 6.6.   Effect of science expenditure on probability of test well success.

in the petroleum industry in which a discovery was made in some zone other than was delineated as the objective. This minimum probability of success can often be estimated based on information from the reserves forecasts for the zone. Many of the forecast methods previously discussed can yield estimates of the number of undiscovered fields remaining by class size. The number for each category can be multiplied by the average area for the category, summed for all categories, and divided by the total unexplored area for the zone to give a crude estimate for the minimum chance of success.

Hopefully, this minimum chance of success can be greatly improved by geological and geophysical effort, shown in figure 6.6 as the sharp incline at the start of the curve. If diminishing returns apply as discussed above, then eventually this incline begins to level off until at some point the incremental gain in success ratio is not justified by the additional science cost. One strategy consideration then is the choice of where on this curve an exploration firm wants to be. Obviously, the preferred level of science effort for one play may not apply for other, geologically different plays. When pretesting strategies for different environments, this difference in preferred science effort should be part of the analysis results.

It can also be hypothesized that additional science effort can increase the average discovery size in a play. This follows from the fact that when prospects are more clearly delineated, it is easier to select those that have larger potential reserves. Again, there would be some economic and technological limit to the effect of science on average discovery size, resulting in a diminishing returns curve similar to figure 6.6.

One possible approach to constructing a curve for the effect of science effort on success ratio or discovery size would be to collect historical data for several firms pursuing differing science programs and relate their success ratios to their average expenditure per prospect. As desirable as such a study would be, it is, of course, impeded by the unavailability of the required data, since they would have to come directly from competing firms. However, it is possible to estimate the success ratios for different firms for different plays from commercially available well data. Well data usually identifies the operator, whether the well was successful or not, and target data, such as location, depth, and objective formation (see Appendix B). While there are certain flaws in this approach caused by limited sample size, inaccurately or incompletely reported data, etc., some useful competitor analysis should result from identifying those firms with the highest historical success. If something is known about the competitors' science programs, their success ratios, coupled with available internal prospect expenditures and internal success ratios, should give an indication of the limits to the curve for the effect of science on success ratio and discovery size.

Most of the remaining environmental and economic variables that are needed to represent the exploration environment can be estimated with greater ease and objectivity than expected discovery size and success ratio. Most of the cost variables are available within the individual firm. The average test well cost and associated variance can be easily related to test well depth using the same empirical curve fitting techniques discussed previously. Similarly, development well costs can be estimated from company data. Some means of modeling the development well schedule is needed in order to properly time the cash flow for economic analysis. Since most companies already employ some means of financially modeling field development for prospect evaluation purposes, a simple solution would be to incorporate the same development scheduling methodology when pretesting exploration strategies. Acreage costs, which may be greatly affected by future exploration outcomes in the area, can at least be estimated at the current prices.

Some other economic variables are determined by the areas in which exploration is to be performed. Customary negotiated deals and concessions including royalty rates, overriding royalties, etc., will need to be considered in order to properly model the economics. This information, too, can often be taken from the company's current prospect evaluation methodology.

The price of oil and other marketing considerations are perhaps more difficult to estimate than expected discoveries. However, price estimates are not likely to be the explorationist's concern because they require a different type of expertise. Instead, price forecasts are likely to be performed either by economists elsewhere in the firm or accepted from external sources. Although the rewards of exploration are surely dependent on the future value of reserves (the selection of the optimum strategy is not likely to be highly sensitive to slight fluctuations in the price of oil but the total revenue will be), the immense body of literature on crude oil pricing precludes the need to include it here. Because the pretesting of an exploration strategy can be done using different price schedules, the sensitivity to price can be investigated and the explorationist can guard against falling victim to erroneous price forecasts.

The explorationist must operate under both physical and financial constraints. Physical constraints include geographic boundaries to exploration areas (or number of prospects), availability of acreage for leasing, availability of drilling rigs, time required for operations, etc. Financial constraints involve the availability and distribution of company resources such as capital and personnel. Incorporation of environmental constraints is critical to exploration strategy evaluations because the first thing that must be determined is whether the strategy under consideration is feasible. Physical constraints are incorporated into strategic modeling as a system of checks and balances on the available "inventory" of resources.

Strategy Composition

The set of alternatives from which a strategy is chosen is composed of decision variables. A decision variable is any choice made by the explorationist or firm that has an effect on the results of the exploration program. Choice of play, science approach, and test well location are examples of decision variables. A specific strategy is some combination of set decision variables.

The first step to building a model that can be used to pretest exploration strategies is to clearly define what composes a specific strategy. Nebulous terms, such as "high science" or "strong acreage position," need to be replaced by decision variables that can be quantified and compared precisely. For example, science effort can be defined by detailing the average geological and geophysical effort to be expended per prospect for a certain play. Once this has been done, the actual cost for the average prospect can be estimated. Similarly, acreage position can be stated in goals for acquiring a predetermined quantity of acreage in specific prospective areas or, simply, capital expenditure for acreage by area. If each decision variable identified can be clearly quantified, and assuming the exploration environment can be accurately simulated, then strategy comparisons can be finely adjusted even for subtle differences.

Obviously it would be impossible (or at least inexpedient) to include every aspect of exploration operations in intimate detail while trying to model exploration strategies. Ideally, the strategy mix should be kept simple while still including the major decision variables. As will be seen, one benefit of strategic modeling is that sensitivity analysis can be done to identify those decision variables that have the greatest impact on the results, so that more consideration can be given to representing them if necessary. With this thought in mind, the following decision variables provide a starting point for modeling exploration strategies.

Choice of basin or play.   Except for smaller firms, this most important decision is usually expressed as a mixture of basins or plays. Depending on the detail of the strategy modeling analysis, this may be done in two steps. First the geographical area (i.e., Permian basin, Gulf of Mexico, Overthrust Belt, etc.) is chosen and must be divided into plays or objective zones. Next the play or mixture of plays is determined. The choice of basins or plays sets the tone for the remaining strategy variables which must be chosen for each play.

Science effort.   If science effort can be successfully related to success ratio as in figure 6.6, then the preferred level, according to criteria such as resulting reserve additions, finding cost, etc., can be found through testing different levels of science expenditure.

Acreage position.   Total acreage holdings can be related to the land budget and prospective area. Because the decision of where to explore was made above, the land budget determines the acreage position. Though partly a decision variable, the exploration budget is constrained by company resources. Furthermore, the budget provides a constraint to other decisions that are dependent on acreage position, such as number of prospects drilling (test well budget) and method of testing. Of course, there exists a physical limitation on total acreage available for the chosen area that sometimes needs to be considered.

Test wells to be drilled.   Once the prospective area has been chosen, the prospects developed through science effort, and the acreage position established, then the number of prospects drilled is constrained primarily by the test well budget. If it were not for this tight company resource constraint, the test well budget would eventually be constrained by the physical limitation of the number of prospects in the chosen area.

Negotiated method of testing.   Often, the explorationist has the option of drilling as operator or farming out prospects. Further complications include farming in when acreage was not available and subtleties such as dry hole contributions. This is a difficult area to represent through simple decision variables, but the main concern is to include applicable constraint systems to insure that strategies are feasible. If constraints on negotiated deals are incorporated, then a check against depending too heavily on outside resources will result.

Risk sharing (working interest).   Like the method of testing, the choice of working interest is affected by outside concerns. In some cases, it may be constrained upward by the unavailability of partners. Other times, it is constrained downward in areas so active that it is nearly impossible to capture all of the acreage for a prospect. Regardless of these ouside influences, working interest is one of the most important decision variables when setting exposure to risk, as was demonstrated in chapter 5.

This list of strategy components may be expanded or reduced depending on the application. However, this step of defining exactly what makes up the set of exploration alternatives and a specific strategy is absolutely necessary for successful strategy modeling (of all types and complexities) and should be done before construction of the exploration model. When the strategy composition has been determined and made quantifiable, attention is turned to constructing a means of pretesting exploration strategies.

Pretesting Exploration Strategies

It was stated previously that the main objective of strategic modeling is to provide support for decision making by the explorationist or exploration management. For complex systems, the most useful form of decision support is the ability to answer "what if" questions about exploration strategies. The explorationist is interested in questions such as: what will our expected cash flow be if we implement this particular strategy? what is the effect on expected reserve additions if we divert expenditure from play 1 to play 2? what if we reduce the expenditure for science and increase the land budget (or vice versa)? what is our chance of program success if we halve our usual working interest and participate in twice as many test wells? and so on. These questions can be answered (or at least examined) only when it is possible to accurately compare different strategies.

One of the best ways to answer "what if" questions about a real world complex system is to build a computer model to simulate the system. The model uses mathematical and logical relationships and elements of the system's inherent randomness necessary to describe the system. An example of a mathematical relationship would be the curve relating success ratio to science effort (fig. 6.6). An example of a logical relationship would be the dependence of development on first having a discovery. Most complex systems also include some elements of randomness. In petroleum exploration, the most obvious elements of randomness are due to uncertain test well results and, when there is a discovery, the size of the deposit. Once these elements or attributes of the system have been identified, they are attached to some framework or structure, often diagrammatical, such as an activity network or decision tree. The model is then tested to insure that it performs as close to the real system as possible within practical limits.

Shown in the most general form, a computer model to pretest exploration strategies would look like figure 6.7. The mechanism or framework for the exploration model stays the same for many areas, however, environmental data from the area under consideration is loaded into the model to give the applicable parameters. Then as various strategies are input, their output results, consisting of return on investment, reserve additions, etc., are compared.

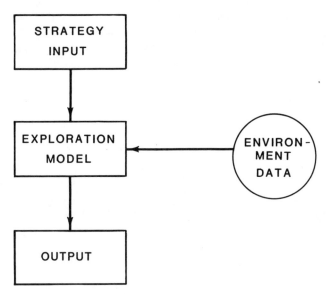

Figure 6.7.   Generalized form of a computer model to pretest exploration strategies.

SOURCE: Reprinted by permission of AAPG from Jones, Buck, and Kelsey 1982.

After the model is completed and implemented on the computer, repetitive passes through the model should reveal the distribution of possible outcomes that can be expected for a given strategy. This type of modeling is often referred to as Monte Carlo simulation because of the repetitive sampling of the random attributes of the system being simulated. The decision variables identified above are then used to perform experiments to answer "what if" questions and eventually arrive at a preferred exploration strategy.

Jones, Buck, and Kelsey [20], in building a model to evaluate exploration strategies in the Permian basin, used a decision tree framework. A decision tree is a network of activities joined by nodes, where a node represents either a conscious decision or a random outcome. Their decision tree for testing prospects can be generalized as shown in figure 6.8. The first activity in the network is to initialize the negotiated deal. This step involves assignment of items such as working and net interests, bonus cost, and method of testing (drill as operator or farmout). Next the prospect is tested or dropped, represented by a node on the diagram because this is a decision. If tested, the test well depth, cost, timing, and other appropriate criteria are determined according to applicable relationships and elements of randomness. The test well may be a success or a dry hole, represented as a node on the diagram since this too is a decision, though not directly controlled. If the test well is successful, the depth and product type (oil or gas) of the field are assigned. Finally, the field parameters, such as reserve size and area of the field, spacing, development schedule, and resulting development costs, are set, again according to strategy specifications and environmental relationships and constraints.

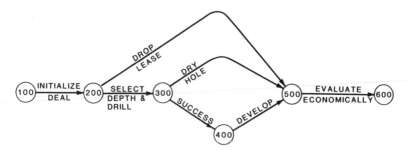

Figure 6.8.   Decision tree network for testing a prospect.

SOURCE: Reprinted by permission of AAPG from Jones, Buck, and Kelsey 1982.

Whether the prospect is dropped or tested, successful or not, it is then evaluated economically to determine its effect on company resources. Note that the decisions in the network are made partly by choice as prescribed by the exploration strategy and partly by random elements in the environment. For example, the decision to drill rather than farmout can often be made by the explorationist but the final test well cost is not known exactly. Conversely, it is not known beforehand that the test well will be successful. However, a conscious decision can be made to improve the expected success ratio by committing more effort to scientifically evaluating prospects prior to drilling.

Figure 6.8 represents testing a single prospect. Hence, it is cycled through for all wells in the program for a given play. This cycle is repeated for the next play and optionally for multiple years in the exploration strategy. The resulting distributions for reserve additions, cash generation, annual production, etc., then give a complete description of the expected results from the strategy being analyzed.

Often, when drilling toward a certain objective, it is necessary to drill through other potentially productive zones, each of which may have its own set of geological attributes. In geologically diverse areas, such as the Permian basin, the effects of such serendipity are financially significant and should not be ignored. Jones, Buck, and Kelsey [20] recognized this significance in their model by expanding the decision tree to four depth intervals, where there could be some chance for success at shallower depths when drilling to a specified depth. Each of these different depth intervals, because they represented different productive zones or even plays, had their own set of reserve estimates performed by the authors of the model for each area of the basin.

In general, the decision tree in figure 6.8 can be expanded for multiple zones as shown in figure 6.9 where zone 2 is beneath zone 1. Note from the diagram that a well drilled in zone 2 may have some positive chance of discovering a deposit in zone 1. If there were a zone 3 deeper than zone 2, then a well drilled to zone 3 may also have some chance of discovering a deposit in either zone 1 or 2, and so on. This interaction allows not only for more realism in simulating the actual exploration process, but for more accuracy in estimating environmental parameters that can then be evaluated by zone.

An important use of the strategy model's output is in analyzing exposure to risk. Each pass through the simulation model represents one possibility for the exploration program's outcome. Some of the passes will show economic return nearly equal to the industry average, a few may show huge returns due to large discoveries (recalling the lognormal, highly skewed distribution of deposit sizes), and some portion of the passes may show no profit at all. If several hundred passes (or possible outcomes) are completed and their results accumulated on a histogram, then the distribution of possible outcomes can be displayed as in figure 6.10. The histogram shows rate of return for all outcomes divided into cells in 10% increments along the horizontal axis. The percent portions of the total outcomes that fell into each cell are shown on the vertical axis.

Risk levels can be estimated with a histogram by summing the percent occurrence for unfavorable outcomes. It is also possible to evaluate the probability of attaining some specific goal or to assess the maximum potential. For the example in figure 6.10, there is about a 14% chance (first two cells) of making little or no profit. But there is a 28% chance (last four cells) of receiving a rate of return greater than 25%. Note that there is even a slight chance of exceeding a 55% rate of return.

The set of simulation outcomes from several passes can also be used to draw a cumulative probability distribution. Figure 6.11 shows the cumulative probability distribution for net company reserve additions as the result of an example explora-

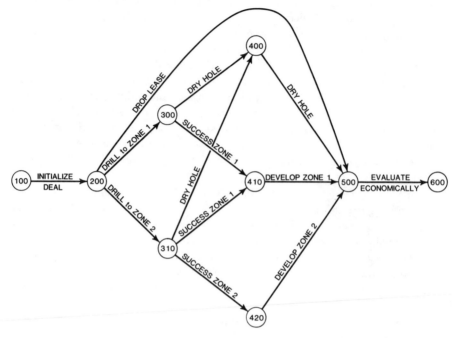

Figure 6.9. Decision tree network expanded for testing a prospect with two potential zones.

tion program. The horizontal axis gives the range of net company reserve additions and the vertical axis gives the probability of adding *at least* the value shown on the horizontal axis. For example, the dashed line shows that there is a 60% chance that at least 900,000 barrels will be added to company reserves if the strategy is undertaken. The tail on the curve shows that there could be as much as four million barrels added. This curve is also quite useful for evaluating the strategy's ability to achieve some set goal. If exploration management had set a goal that said that at least 500,000 barrels should be added as a result of the exploration program, then this particular strategy would be assessed as having an 80% chance of achieving that goal.

Graphs of distributions of probabilistic outcomes have several advantages over the usual single point estimate of average or expected return as has already been seen. Another important use of these graphs is in adjusting exposure to risk. Figure 6.10 gives about a 14% chance of program failure. If exploration management decided that this percentage constituted too much risk, then another strategy, perhaps one that had a lower average working interest but in more wells, could be tested and evaluated in a like manner. Note that in such a case, no difference would be noted in the expected rate of return. Hence, the expected value analysis would

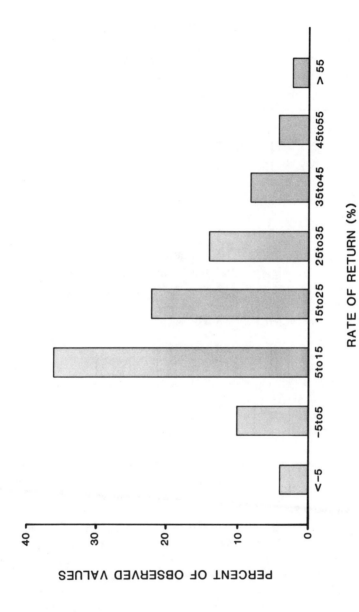

Figure 6.10. Example histogram of simulated exploration program outcomes.

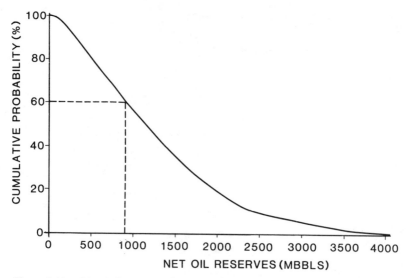

Figure 6.11.   Cumulative reserves distribution for net reserves additions as the result of an example exploration program.

SOURCE: Reprinted by permission of AAPG from Jones, Buck, and Kelsey 1982.

show no difference. However, the probability that the program would show no return would be greatly reduced, as would be indicated in the histogram or cumulative probability graph. An important function of the strategy model then is to aid the explorationist in minimizing risk or setting the preferred risk level both at the local level and integrated with corporate strategic objectives, as was discussed in chapter 5.

If some variable from the strategy or environment is selected and varied upward and downward, it is possible to see what effect, if any, that variable has on the results of the exploration program. This type of analysis is commonly called sensitivity analysis. Sensitivity analysis is useful when building the computer model because it identifies those variables, both in the environment and in the strategy, that need special attention or accuracy because they heavily influence the performance of the model (and the performance of the strategy in reality). For example, the production schedule for a given field, if altered slightly, such as by raising the production rate early in the well history, has a marked effect on the cash flow and, hence, on the rate of return, even if the total cumulative production is unchanged. Therefore, in this example, care should be taken to see that the production schedule is accurately modeled in the computer representation.

More important, once the computer model has been completed, sensitivity analysis will point out what effect the decision variables have on the economic and reserve outcome of alternative strategies. Generally, this is done by making several strategy evaluations, each time varying the strategy component of interest slightly while holding other components constant at some base case. Whatever output figure is of interest, such as reserve additions, rate of return, net present value, etc., is then plotted against the change in the strategy variable. For example, recall that earlier it was hypothesized that the effect of science on success ratio and discovery size showed great improvement followed by diminishing returns. Because this suggests that past some point, additional science effort becomes uneconomical, sensitivity analysis should reveal the desirable science level for the area under consideration. Figure 6.12 shows a graph of the effect of science expenditure on rate of return for an example exploration program. The rate of return increases with additional science up to a maximum at the preferred expenditure level, past which the additional cost does not appear warranted.

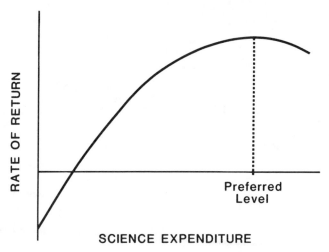

Figure 6.12.   Example sensitivity analysis of science expenditures.

An important point to keep in mind when performing this type of sensitivity analysis is that several measures of merit should be used. For example, in figure 6.12, the preferred science level was based on rate of return. But the graph would look quite different if reserve additions were shown instead of rate of return, because additional science effort surely would never reduce reserve additions if all other strategy components remained constant. Still another preferred level might be set if net present value was used instead of rate of return or reserve additions.

The great flexibility of analysis approaches provided by strategic modeling should be exploited to insure that a gain in one direction is not a loss in another.

One possible pitfall that must be avoided in strategic modeling is the choice of a consumptive strategy. A consumptive strategy is one that shows great short-term profit at the expense of reducing the firm's position for future exploration. For example, a strategy that diverts all funds to testing prospects on the shelf and none to acquiring new acreage and developing new prospects would probably show great short-term return but would eventually leave the firm with no future in the area. While this may be an actual objective when pulling out of a play that no longer proves profitable, more often such a strategy would be devastating to the firm's longevity.

The easiest way to avoid this pitfall is to have the strategy model maintain a simple inventory of acreage and prospects in various stages of development. As a strategy is executed, tested prospects are removed from the inventory of new prospects while acreage and science expenditures add to the inventory. At the end of the simulated exploration program, the ending position can be scrutinized. In some cases a strategy may have to be adjusted because it consumes too much of the firm's position in the area. Another strategy may have to be adjusted because acreage is expiring untested, and so on. In this way, strategic modeling can be equally concerned with short-term economic goals and long-term growth.

Jones, Buck, and Kelsey [20] describe in detail a strategy model developed for use in the Permian basin of west Texas and southeast New Mexico. They divided the basin into six geologically similar areas and divided each into four depth intervals. They forecast average discovery size by area by depth interval based on the discovery process model developed by Drew, Schuenemeyer, and Root [12] discussed before. Jones et al. went on to use these forecasts in an exploration strategy model and discuss the uses and output of such a model. They discuss the comparison of alternate strategies using the model and provide an example based on actual environmental and strategic data. McKinney and Jones [21] give another example of detailed exploration strategy modeling. Their international exploration strategy model consists of three main submodels for the spatial occurrence of deposits, the physical discovery process, and exploration economics.

Because the use of detailed modeling for exploration strategies is a rather new field, it cannot be approached by any current, easily accessible formula. The development of such methodology within the firm requires a combination of past experience and new thinking. Like all forms of evaluative and planning methodologies, it requires cooperation and collaboration between exploration management and the planning and evaluation functions within the firm.

Obviously, an exploration strategy model is not developed to replace human decision, for which there is no substitute. Instead it supplies an evaluative tool, an analysis environment or "wind tunnel" test of operations too complex and expensive to experiment with directly. The process is usually iterative with the model constantly evolving and changing when necessary to represent a new area or nuance of exploration; with its capabilities exploited and its limitations observed. And while the type of detailed modeling discussed in this chapter involves more front-end development time and coordination between functions, the potential of such modeling lies in its extreme comprehensiveness and flexibility.

References

1. Hubbert, M.K.: *Energy Resources, A Report to the Committee on National Resources,* National Academy of Sciences – National Research Council Pub. 1000-D (1962). Reprinted 1973 by National Technical Information Services, U.S. Dept. of Commerce, No. PB-222401/2.

2. Hubbert, M.K.: *U.S. Energy Resources, A Review as of 1972,* Serial No. 93-40 (92-75), U.S. Government Printing Office, Washington, D.C. (1974).

3. Wiorkowski, J.J.: "Estimation of Oil and Natural Gas Reserves Using Historical Data Series: A Critical Review," unpublished manuscript (1975).

4. Mayer, L.S., Silverman, B., Zeger, S.L., and Bruce, A.G.: "Modeling the Rates of Domestic Crude Oil Discovery and Production," Report to the Office of Information Validation, Energy Information Administration, Princeton University, Dept. of Statistics (1979).

5. Schuenemeyer, J.H.: "Comment" (to Wiorkowski, 1981, below), *J. of Amer. Statistical Assoc.,* Sept. (1981) 76, no. 375, 554-558.

6. Wiorkowski, J.J.: "Estimating Volumes of Remaining Fossil Fuel Resources: A Critical Review," *J. of Amer. Statistical Assoc.* (Sept. 1981) 76, no. 375, 534-548.

7. Kaufman, G.M., Balcer, Y., and Kruyt, D.: "A Probabilistic Model of Oil and Gas Discovery," *Methods of Estimating the Volume of Undiscovered Oil and Gas Reserves,* J.D. Haun (ed.), AAPG, Tulsa, OK (1975).

8. Kaufman, G.M.: *Statistical Decision and Related Techniques in Oil and Gas Exploration,* Prentice Hall Inc., Englewood Cliffs, NJ (1963).

9. Barouch, E. and Kaufman, G.M.: "Probabilistic Modeling of Oil and Gas Discovery," *Energy: Mathematics and Models,* F.S. Roberts (ed.), SIAM, Philadelphia, (1976).

10. Arps, J.J. and Roberts, T.G.: "Economics of Drilling for Cretaceous Oil on East Flank of Denver-Julesburg Basin," *Bull.* AAPG (Nov. 1958) 42, no. 11, 2549-2566.

11. Schuenemeyer, J.H. and Root, D.H.: "Computational Aspects of a Probabilistic Oil Discovery Model," *Proc.,* Amer. Statistical Assoc., Statistical Computing Section, (1977) 347-351.

12. Drew, L.J., Schuenemeyer, J.H., and Root, D.H.: "Petroleum Resource Appraisal and Discovery Rate Forecasting in Partially Explored Regions; Part A, An Application to the Denver Basin," 1138 USGS (1980).

13. Singer, D.A. and Drew, L.J.: "The Area of Influence of an Exploratory Hole," *Economic Geology* (May 1971) 71, no. 3, 642-647.

14. Schuenemeyer, J.H. and Drew, L.J.: "An Exploratory Drilling Exhaustion Sequence Plot Program" *Computers & Geosciences* (1977) 3, no. 4, 617-631.

15. Hendricks, T.A.: "Estimating Resources of Crude Oil and Natural Gas in Inadequately Explored Areas," *Methods of Estimating the Volume of Undiscovered Oil and Gas Reserves,* J.D. Haun (ed.), AAPG, Tulsa, OK (1975).

16. Jones, R.W.: "A Quantitative Geologic Approach to Prediction of Petroleum Resources," *Methods of Estimating the Volume of Undiscovered Oil and Gas Reserves,* J.D. Haun (ed.), AAPG, Tulsa, OK (1975).

17. Fisher, F.M.: *Supply Costs in the United States Petroleum Industry*, Johns Hopkins University Press, Baltimore (1964).
18. Erickson, E.W. and Spann, R.M.: "Supply Response in a Regulated Industry: The Case of Natural Gas," *Bell J. of Economics and Management Science* (1971) 2, 94–121.
19. MacAvoy, P.W. and Pindyck, R.S.: *The Economics of the Natural Gas Shortage (1960–1980)*, Elsevier North Holland, Amsterdam (1975).
20. Jones, D.A., Buck, N.A., and Kelsey, J.H.: "Model to Evaluate Exploration Strategies," *Bull.*, AAPG (1982) 66, no. 3, 317–331.
21. McKinney, G.W. and Jones, D.A.: "Model to Evaluate International Petroleum Exploration Strategies," paper presented at the TIMS/ORSA Joint National Meeting, April 25–27, 1983, Chicago, IL.

7

Avoiding
Planning Pitfalls

Introduction

Some of the common planning pitfalls are discussed in this chapter. The size and sophistication of the organization directly determines the number and type of pitfalls that the planning effort may encounter. The primary problem common to organizations of all sizes is that planning efforts cannot succeed if top management does not see the need for planning. The CEO must be committed to the planning effort and his support must be known to all of the line managers involved if the planning effort is to be successful. A formal planning organization is not necessary in order to plan. Planning, it must be remembered, is a managerial responsibility, and the people who are called "planners," are merely facilitators. In many large companies with large formal planning departments, management tends to delegate a part of the executive planning responsibility to the planning staff. This is, in most cases, a mistake, since direct input from top management is needed for planning to be successful.

The key problem areas discussed in this chapter are: (1) the use of forecasting as planning or in lieu of planning, (2) risk analysis approaches and their possible problems, and (3) communication problems in planning.

Forecasting Versus Planning

Forecasting Syndrome

As discussed in chapter 2, the process of forecasting is often confused with planning. Forecasts can be used as a basis for different actions within the planning process, but they do not constitute planning. "Best guess" forecasts, when used alone, are almost always wrong. If a plan is based only on a single scenario for each strategic business unit (SBU), it most likely becomes a forecast and not a plan. A forecast, which assumes history will repeat itself and, thus, is based solely on historical performance without considering future events, is not a plan. Plans should be based on certain conditions and desired results while forecasts deal with projections without knowing the final result. The original disciplines of planners or managers have definite influence on their planning approaches. Most accountants and engineers are taught from the beginning to deal with certainties and tend to look at planning as merely forecasting. They are most comfortable with deterministic alternatives and rely heavily on historical data.

Forecasting Limitation

Many companies do no real planning because what they call planning is merely forecasting. Such forecasts are merely projections of future capital and expense resulting in future cash flows. Scenarios or strategies are either nonexistent or purposely vague in these companies. They make no attempt to state time frames for goals, which are too often steeped in catch all type objectives.

The most important phase of planning is the actual planning itself, which requires consideration from top management about the firm's current and desired future position. This forethought cannot be done by the planning staff subordinates. Forecasts or projections can be done without this forethought, but the results are not real planning. Planners can use a trial and error method in an attempt to get the answers that they think fit management's concepts. This simulated approach to planning is done by manipulating the capital allocations to arrive at exploration plans. Another common "planning" method uses historical factors, such as finding costs and the relationship to reserves found per dollar spent, to forecast future results. Neither of these methods considers exploration strategies or controls where prospects will be drilled.

Numerical optimization based on economic analysis is another approach used to establish exploration capital allocation plans. This approach, though financially sound, induces important practical problems. During the generation of a prospect, where there is little known about it, input to the economic analysis is based on little or no fact. Because of optimism, the economics are generally overstated. As the detailed geology and geophysics are worked out on a prospect and more is known about it, the economic analysis becomes more realistic. In such a case, the optimization method results in the company funding all the immature prospects with "higher" reserves and better economics. This is sure to be undesirable.

Proper Use of Risk Analysis in Planning

Over the last decade the term "risk analysis" has been much used. Probability theory, decision trees, and Monte Carlo simulation are all terms used in the broad umbrella of risk analysis or decision theory. Newendorp [1], Megill [2], Cozzolino [3], Grayson [4], and McCray [5] have published full length works on risk analysis. applications for the petroleum industry. Chapters 5 and 6 expressly address some risk analysis techniques and cover the technical pitfalls of risk analysis. However, problems can be introduced through subjectivity.

In most cases, the analysts who are performing prospect risk analysis and resulting economic projections must often rely on subjective probabilities in their calculations. Subjective probabilities, of course, are not always realistic. Geologists are by nature perpetual optimists. They are always sure that they can beat the odds and find that one remaining giant field left in the basin. Therefore, they have a tendency to assign unrealistic chances of success of finding the large size fields. If an exploration program is described by a series of expected values where the future field sizes and chance of success are exaggerated, the resulting economic projections are distorted. The cash flow generated as a result and the accompanying projected reserve additions are unrealistically high compared with those that are based on historical data of actual reserves found.

If, in the planning process, the subjective probability assignment methods are used in the risk analysis, where historically based data is lacking, it is necessary to have these probabilities screened by someone who has the experience to make judgments and adjust the subjective probabilities so that realism is achieved and uniform results are obtained. If remaining field size distributions cannot be forecast directly from historical data, the distributions that are used need to be carefully screened. The use of expected value data in the planning process, whether used in portfolio selection or in future cash projections, needs to be tempered with judgment and experience.

If the future field sizes and their distribution are not easily determined from historical basin analysis data, a Monte Carlo simulation model approach may be used. Newendorp [1] is strongly in favor of models using separate distributions to describe input variables such as pay thickness, productive areas, and fill up. In this manner, the explorationist can relate distributions directly to geological processes. Most models of this type allow for various types and shapes of distributions to be used for describing these variables. From experience, geologists seem to prefer triangular distributions. However, triangular distributions are very poor estimators for the lognormal, highly skewed distributions found in petoleum geology. Megill [2] discusses advantages and disadvantages of triangular distributions in his work on risk analysis. Depending on the type of computer models used, any number of variables can be used as input. This input can come from a number of different sources so that the expertise of the entire firm can be used.

If a basin or area is virgin or very immature but the large structures have been identified, the simulation model approach can be used to make what Newendorp [1] refers to as the basin reserve model. This model can be run using minimally available data in the form of distributions. The basin reserve model then emulates drilling, using different scenarios to sequence the drilling and to determine other controlling factors. The output can then give distributions of results and can be used to perform sensitivity analysis of various input distribution parameters. In new exploration areas, where costs are important but not specifically known or easily controlled, the effects of cost variance can be estimated by Monte Carlo simulation.

Some recommendations that Megill [2] gives are: (1) isolate key variables by analyzing data as they are gathered, (2) quantify the positive and negative effects that these key variables have on decisions, and (3) insure that uncertainty is handled at the input variable level and that the input is not already adjusted for risk. Megill's final recommendation is that answers be given as sets of cumulative distributions instead of stand-alone expected values. According to Megill:

> *Exploration is a process which commits company funds to an unknown future. The unknowns involve not only geologic uncertainty but a number of critical economic factors such as price, cost, inflation, and possible changes in tax laws. Exploratory funds, therefore, are committed to a sum of unknowable expectations — not to facts.*

This statement is not only applicable to risk analysis alone but to all strategic planning for petroleum exploration management.

Communication Gap in Planning

The most common cause of planning failure is simple communication breakdown. Failure to communicate can be in both the written word and spoken word. Most communications fail because of the human assumption that other people know basically what we are talking about and that we are merely filling them in with some pertinent details. We are negligent in finding a satisfactory level at which both parties are of equal understanding and where basic communications can start. If we do not begin our communication at some point where everyone involved has common understanding, we cannot expect to succeed at communicating.

Planning Guidelines

All formal planning processes start with a set of ground rules, often referred to as guidelines. These guidelines usually are imposed by top management in the organization. In large companies, planning guidelines may cover several operational levels in detail and may have undergone several modifications before reaching the operating level. Planning guidelines set forth the mechanics and philosophies of planning. Some specifications, such as the product prices to be used based on an expert's forecasting, would seem straightforward. However, pricing specifications should include the economic assumptions that were made to arrive at the price projections. If the guideline assumptions were not communicated properly, the managers putting together the exploration programs may make economic assumptions in their planning that are contrary to the ones used by the experts that established the pricing forecast. The scenarios being examined should be carefully explained in detail to operational managers so that they know what restrictions they need to follow and what objectives and goals are expected of them.

Guidelines seldom fully cover all the aspects that need to be understood. One method for achieving better understanding is to have preliminary guidelines issued to managers and planners involved and followed by a meeting where the guidelines are discussed in detail. The final guidelines that are issued after this dialogue usually are much more manageable and are more easily understood by everyone involved. The alternative method to this approach is to have the operating entities ask questions about the original guidelines, which generally results in clarifying amendments to the guidelines. The amendment method still allows too much interpretation and misassumption, which may cause discontinuity in the finished plan.

The Planner's Role

As mentioned earlier, the main function of a formal planner at any level is facilitation. This requires the planner to be a major communicator in the organization. The planner is responsible for initiating the planning process. In this capacity, the planner provides the informational input and coordinates the start-up of the planning process. An additional function of the formal planner is that of reviewer and evaluator of the plans prepared. The planner reviews and evaluates the consistency of plans made by line managers and consolidates these divisional plans to make sure that they are consistent with the stated mission and strategies of the company. In the case of exploration planning, the planner should help coordinate the business strategies with exploration strategies in the plays that constitute the expression of these strategies.

Obviously, all of these functions require that the planner be an effective communicator. If the planner does not communicate his ideas and the ideas of upper management to the operational level, the planning process will be hindered and may fail. The planner also has to be diplomatic in order to get management at all levels to cooperate in the planning process.

Too often planners have dreams of grandeur and forget their place in the planning scheme. If a planner tries to control the planning process too closely in order to turn out an elaborate planning document, he becomes part of the problem. Planners must remember that the formal plan is secondary to the real planning process.

The thoughts and dialogue that occur in the process are what guides the company, not the formal planning documents.

The Managerial Role

It needs to be stressed again that the CEO and top line managers *must* take an active part in the planning process, and they must set the direction and emphasis of the plan. In addition, they must communicate their endorsement of the planning process to their line or operational managers. If this is not done, the line and operational managers, who are busy with operational problems, will view the process as mere busy work that interferes with their day-to-day jobs. Many operational managers are acquainted with budgets or short-range operational planning but are not acquainted with strategic planning concepts. The purpose of this book is to get exploration management or operational management integrated with the strategic planning approach.

Larger firms have more levels of management that must be reached by planning communications and, hence, have a greater likelihood for communication breakdown. Traditionally, large integrated and diversified firms are divided into operating groups or strategic business units (SBUs) with many formal planning systems in effect. The SBU staff must generate their own plans, including SBU goals, objectives, and strategies, which are then presented to corporate management where a massaging and review process eventually results in the corporately blessed strategic plan. Kiechel [6] is critical of this form of "canned" strategic planning. His main criticism is that strategies that are created in this fashion are almost impossible to implement. The reason for this problem seems fairly obvious. The strategies are not germane to the operation's objectives and goals and to the corporation as a whole. These strategies are generated by the SBU staff and not by the operating or line managers who should determine the operating strategies. Sawhill [7] views this rejection of strategic planning as the rejection of a single strategic viewpoint that is based on forecasting and not on planning. Sawhill contends that many past mistakes in corporate strategy generation were caused by attempts to create certainty instead of developing flexible strategies to handle the uncertainty found in energy exploration. He asserts that strategic planning approaches in the future will require pragmatic market and competitive assessment coupled with creative and flexible approaches to planning.

Most of the communication problems and related pitfalls discussed here point to the fact that planning is a very complex process that requires considerable thought by management. Resulting descriptions of planning concepts require that detailed explanations be communicated to all line managers so that continuity can be maintained. These observations suggest that the best way to maintain continuity and consistency is with a combined "top down" and "bottom up" planning system.

References

1. Newendorp, P.D.: *Decision Analysis for Petroleum Exploration*, Petroleum Publishing Co., Tulsa, OK (1975).
2. Megill, R.E.: *An Introduction to Risk Analysis*, Petroleum Publishing Co., Tulsa, OK (1977).
3. Cozzolino, J.M.: *Management of Oil and Gas Exploration Risk*, Cozzolino Associates Inc., W. Berlin, NJ (1977).
4. Grayson, C.J., Jr.: *Decision Under Uncertainty*, Boston: Division of Research, Harvard University, Cambridge, MA (1960).
5. McCray, A.W.: *Petroleum Evaluations and Economic Decisions*, Prentice Hall Inc., Englewood Cliffs, NJ (1975).
6. Kiechel, W.: "Corporate Strategists Under Fire," *Fortune* (Dec. 17, 1982) 106, no. 13, 34–39.
7. Sawhill, J.C.: "How to Rescue Strategic Planning in Energy," *Oil & Gas Journal* (Feb. 28, 1983) 81, no. 9, 90–92.

8

Conclusions

Strategic planning for petroleum exploration management must include a multitude of topics from both the exploration process and the strategic planning process. This chapter briefly summarizes and links together several previously mentioned topics important to strategic planning.

Scenario Generation

A detailed discussion of corporate scenario generation for large diversified or integrated petroleum companies is beyond the scope intended here, but a general discussion of how scenario generation would be applied in exploration and production companies is given.

The purpose of a scenario is to plausibly describe future events. Depending on the imagination of the planner and the intended use of the scenario, many kinds of scenarios can be constructed. However, a scenario must be internally consistent. The scenario must have a central theme or core that is descriptive of real events that lead to possible future circumstances. The future described by the scenario need not be desirable to the firm. The choice of scenarios is usually determined by developments outside the control of the company. In some cases, scenarios will be based on external contingencies (for example, if crude prices drop below $25 a barrel). Often, organizations attempt to choose a scenario on which to base their plan, which is usually impossible. Scenarios are only used to decide how the firm should react to specific external events. It is these responses that create strategies. Strategies are not part of the actual scenario.

Scenarios are not forecasts and, hence, are not necessarily right or wrong, as long as each scenario is internally consistent. Scenario generation is a decision-making tool for management. Scenarios provide a method for developing strategies to answer "what if" questions and for developing contingency plans. However, many companies make the mistake of trying to make their scenarios too specific. It must be recognized that today's trends and events often are not the most important scenario elements. Scenario writing is a skill developed only through practice. The writer of scenarios soon learns that simply identifying possible trends and events is not as important as describing the logic behind them. It is the logic behind trends that leads to future events. This logic needs to proceed from basic assumptions. Detailed open discussions of assumptions and trends or events will test the validity of the scenario.

The scenarios should be conceived by the executives of the corporation, who can identify the many factors that may affect the organization. Scenario generation should not be entirely delegated to staff employees unless they have the petroleum experience necessary to truly judge the effects of variables in the environment. Each scenario must be selective enough to bear directly on the issue under consideration. Hence, many scenarios may be needed to represent the many company choices and possible circumstances in the future.

There are a variety of ways to write scenarios. Some ways are straightforward while others are complex requiring large amounts of documentation. Others may be impressionistic, depending on what factors are being covered. For instance, world affairs will impact crude prices, affecting small independent companies as well as large multinational companies, although in different ways. In petroleum exploration, technology also has a direct impact, which varies for companies of different size. For example, large companies can directly benefit from new scientific techniques that may be too expensive for small companies.

When developing a scenario, the decision-making parameters must be carefully defined. For an exploration company, this is rather simple compared with other businesses. Next, the starting point of the scenario, that is, present conditions, must be defined. The next step is to characterize the trends and external pressures that will affect the scenario. These influences often include a number of sociological issues that affect the industry. The last step is to determine the extent and direction of the scenario or group of scenarios. This is best done by addressing "what if" questions. Scenarios may extend in various directions depending upon these "what if" questions about influences like government regulation of prices or changes in product demand.

Scenario writing requires a wide range of assumptions based on trends and underlying theories that bind the scenario together. Scenario validity is dependent upon the quality and realism of the available information. When scenarios must postulate political reactions and economic responses to events, such as Middle East unrest and its effects on crude supplies, the possible chain of assumed events must be explicitly examined.

As was mentioned earlier, planning is not intended to eliminate risk altogether but rather to select the risk to be taken. Scenario generation can be used to identify some of the risks that might occur in the future. For instance, if a planner for an oil company had identified early trends in 1980 or early 1981 indicating the possible occurrence of a world oil glut in 1982, he might have set actions in motion to minimize the effect of crude oil price reductions on his company. There were a few companies that actually did see this possibility of an oil glut and took measures, such as selling off producing properties, to pay off debts and improve their cash position.

There are various opinions in planning literature about the value of planning with multiple scenarios under conditions of uncertainty. This type of approach does give some methods to help identify future problem areas and to help define "trigger points" for contingency or alternative planning. This does not mean that there should be an attempt to limit the number of scenarios. One approach is to use three or four scenarios, giving bounds to the limits of uncertainty. These scenarios help determine the middle ground for the "best guess" scenario. An important drawback of the best guess scenario is that internally it will probably be more complicated than the extreme limit or bounding scenarios. The best guess scenario has to include a larger number of trends and therefore will be the most complex. Using a single best guess scenario is not the safest planning method. Such a singleminded method would suppress alternatives and therefore suppress imaginative planning. Also, a scenario is not a true forecast and it is very unlikely that the best guess scenario will truly be indicative of the future. Scenarios only provide a framework for considering changing trends and industry events and are a key element in the planning process.

In times of uncertainty, multiple scenarios provide a good basis for contingency planning and allow the monitoring of early warning signs that provide flexibility to the planning and decision processes. Depending upon the size and scope of a company, the number and types of scenarios should be dependent upon the details that top management wants and how much time and money they are willing to invest, because they are the ones that will use the possible outcomes in their decision making. Again, scenarios are just tools for decision making in an uncertain environment.

Some of the external factors that should be considered in scenario generation are:

1. future economic environments in the United States and the world;
2. future crude oil supplies, United States and foreign;
3. future demands for gas, including terms of future gas contracts and decontrol;
4. future inflation and corresponding interest rates;
5. future change in direction of federal government regulations; and
6. use of alternative fuels in the future.

Attainable Objectives, Goals, and Strategies
for Operational Management

One of the prime purposes for a formal planning system is to develop methods and measures to insure that goals and objectives established are realistic and attainable. This must be done for two obvious reasons. First, in order to make the plan more than just an ignored document, the operational goals and operational objectives set forth in the plan must be attainable by operational management. Second, the performance measurement and control system requires that these goals and objectives be used as the basis for measurement of operational management. If these goals and objectives are not attainable by operational management, the performance system is meaningless.

There are two classical planning methods that can be used to set these objectives and goals, the "top down" approach and "bottom up" approach. In the top down approach, the operational goals and objectives are assigned by top management to the operating levels of management by edict. This approach usually guarantees that operational goals and objectives will be consistent with the corporate mission, business objectives, and strategies. However, this does not necessarily mean that the operational level can attain their objectives and goals. Corporate management usually does not know about operational problems that might prevent attainment of the top down goals. The bottom up approach gives operational management the option of describing operational objectives and setting operational goals according to operational implementation of corporate mission and business objectives, goals, and strategies. With the bottom up approach, operational management is expected to understand the corporate position but many times they are too far removed from the corporate areas, especially in large companies, to really interpret corporate problems. There are obvious problems with both approaches.

One method that takes the best from both top down and bottom up approaches is called "goal factoring" or "goal negotiating." In this planning method there is a negotiation process between operational management and corporate group management. The methodology and approaches that have been described in this book are aimed at providing goal acceptance at all levels for many reasons. Goal acceptance is more likely if participation is from all levels. The authors recommend using the portfolio approach to give management a selection of opportunities of use to corporate management for incorporation of exploration programs. The corporate portfolio is composed of the collected objectives and goals from each of the operating groups. The portfolio approach is used to combine both the exploration business and other business alternatives. To quote Ben Ball [1]:

> *It is not the corporate objectives and goals that should be communicated to the operational level but rather the objectives and goals that have been determined to be ap-*

propriate for that particular operation, *that defines to a large degree their role in the corporation's portfolio.*

This method defines goals and objectives in a manner agreeable to both corporate and operational management. Abouzeid [2] gave results of a survey of 200 companies in which the process of goal identification and achievements was reviewed. From the survey, he synthesized a model that simulates and formalizes the planning actions previously discussed, including the performance review. Abouzeid refers to the performance review as goal implementation in this stage of his model. Successful goal implementation relies on the control and incentive system.

Controls and Incentive Systems

In chapter 2 the control system was discussed. The management reporting system was described as three pronged: the key factor reporting system, the operating intelligence system, and the environment scanning system. In Newman's [3] breakdown of this control system there are three parts: steering controls, yes-no controls, and postaction control. Steering controls are probably the most important because they prevent harmful actions or decisions that will cause the organization problems in the future. To provide the steering controls, Roush and Ball [4] developed the critical implementation factor (CIF). Their method is designed to monitor progress toward a desired strategy. The CIFs are geared toward the four critical implementation areas of organizational structure, human resources, corporate culture, and control systems. Critical implementation factors serve a dual purpose. First, they check to see if the strategy is still valid, and second, they check performance criteria for validity. The CIF concept is difficult to implement because these implementation factors, to be effective, have to be controlled directly by the chief planner, the CEO.

The critical implementation factor is just part of the total problem discussed by Roush and Ball [4] concerning the methods used to design a control system that will insure strategy implementation. Kiechel [5] criticizes traditional strategic planning and very pointedly singles out nonimplementation as the problem with strategic planning. Roush and Ball's CIFs appeared two years before Kiechel's ariticle. It is obvious they recognized the problem earlier and suggested a solution through the control system. Their four critical organizational factors are interrelated with the control systems and have direct bearing on efforts to implement strategies. The control system should be of primary importance and should monitor the other organizational factors as they effect the strategy implementation.

Roush and Ball [4] identify six steps or processes in the strategic control system:

1. formulate strategy and specify strategic objectives
2. evaluate strategy against implementation factors
3. specify performance criteria for strategy achievement
4. develop measurement methods
5. develop a reporting process
6. specify corrective actions

Recalling the four planning phases discussed in chapter 2, step 1 must be done very early in the planning process in the formulation phase. Step 2 occurs in the implementation phase where it is tested against the four organizational factors of organizational structure, human resources, corporate culture, and control. Step 3 occurs in the organization and control phase and requires the definition of key success requirements implied by the strategy. Step 4 also occurs in the organizational and control phase and is used in more conventional control systems. Step 5 is part of the management information system previously discussed. Step 6 is done in the reformulation phase. Roush and Ball emphasize that this system should be carefully tailored to the CEO's requirements and made as simple as possible. This strategic control system seems straightforward but will take dedication and effort by the CEO and other top management to make it work.

Often, the lack of strategic implementation is blamed on operating management. However, operating groups are bound by conventional control systems where they are measured on performance of short-term operating results. These short-term measures often look good but may not be. For example, in an exploration and production operation, monthly cash flow or earnings will look better if no test well drilling is done because such items as dry hole writeoffs have direct negative effect on earnings. If this is carried to extremes and no test wells are drilled then no reserves will have been found. The operation is making short-term profits at the expense of staying a viable business. This is the consumptive strategy discussed in chapter 6. Exploration management must not be measured on short-term results if an exploration strategy consistent with the organization business strategies is to be maintained. As Ben Ball points out in his article discussing planning and operational management [6], the control system has to be designed so that short-range controls are imbedded in long-term objectives and goals. This way, operational people will conduct exploration and production operations in a fashion consistent with long-term strategic planning. This cannot be done except with the cooperation of the CEO or top management team.

The reward or incentive system, which is part of the overall control system, provides a method of rewarding the key managers that are in positions responsible for implementing the operational plan including implementing the strategy. A detailed discussion of incentive systems is beyond the scope intended here. However, the incentive system, which provides motivation, is an important part of the control system and should tie into the strategic planning system and the implementation of the strategic planning process. Batten [7] devotes a chapter of his management text to motivation and describes motivating factors other than monetary incentives. In the previous discussion of planning, many areas that run parallel to Batten's premises were discussed, including the various aspects of corporate culture and human resources. Management can easily devise a bonus system, stock option, or other monetary rewards for specific action, although great care should be taken that these monetary rewards are based on long-term objectives and goals and not on short-term operating results. This monetary incentive approach has a definite place in the system; however, there are nonmonetary incentives that have long-term effects. For example, developing a corporate value system that matches those of managers and employees is highly effective. If there is commonality of purpose between the manager's personal goals and objectives and those of the firm, then the likelihood of successful implementation of planning and ultimate success of the company reaching its goals and objectives are much improved. In some cases, strong measures will be needed to modify the present corporate culture to a culture that is compatible with the aspirations of the company as envisioned through their strategic plan.

Flexibility of Strategies

In times of uncertainty, flexibility in the planning process is absolutely necessary for survival. The use of scenario generation is one method that can be used to consider a set of possible future environments. The possible responses to certain events are strategies. The scenario generation approach will give a set of strategies that will fit a series of events and future environments. These contingent strategies have a direct effect on both business and exploration strategies. For instance, if a scenario had been predicted several years ago in which an oversupply of natural gas was a possible event, then an accompanying exploration strategy could have evolved that could have been used when early indications of this event first surfaced in 1981.

Another planning tool is the use of contingency planning. This approach was mentioned earlier. The advantages of contingency planning over a single scenario plan are that it gives flexibility to the plan and to the accompanying strategies, determining the trigger points at which each of the contingency plans are implemented. Contingency planning requires a full understanding of successions of events and the strategies that are needed to deal with these events. Also, trigger or decision points have to be indicated directly in the control system. These trigger points need to key off early indicators of the actual events. With contingency planning, strategies can be devised in advance for the contingency plan and then the accompanying exploration strategies can also be devised in advance. In many cases, the exploration strategies will have to be devised in a fashion that enables direction and emphasis to be changed without completely disrupting the exploration effort. This type of contingency planning adds valuable flexibility to the planning process. Sawhill [8] emphasizes the need for flexibility in strategic planning for energy. According to Sawhill:

> *Planners must become the advocates for flexibility, making a forceful case for the capital investments and organizational changes needed today to respond quickly to the unexpected events of tomorrow. Unfortunately, a short-sighted reaction of many inexperienced managers to a hostile and uncertain environment is to throw out planning and slash all costs to the bone. But by doing so they frequently destroy the very flexibility they need to cope with change.*

Summation

This book should provide an understanding of the strategic planning process for exploration management so that the exploration operations can be conducted as business units. By using some of the methodology and techniques suggested, management can examine exploration from a strategic business standpoint. Exploration management must be able to convey to corporate management the varieties of strategies and options available to them. This means that the operational group must be able to define their strategies in specific business terms including viable risk management. This will give corporate or group management a portfolio of the various available exploration strategies. Once corporate management has chosen the strategy or strategies that exploration should follow, it is then up to corporate management to allocate the resources necessary to pursue the strategies. Corporate management must develop policies for the long-term business strategies, objectives, and goals expected of exploration operations based on the operational objectives and goals. It should be reemphasized that both the business strategies and the resulting exploration strategies must be flexible and that measures must be taken to insure the implementation of these strategies.

John Sawhill concluded his article on strategic planning [8] with a thought central to the topic of this book:

> *If strategic planning is like doing a crossword puzzle, a challenging and praiseworthy process that is ultimately useless, it deserves to be abandoned. But future challenges in the energy industry require pragmatic market and competitive assessments, creative approaches, and flexibility — a broader, more effective approach to planning than that of the past.*

References

1. Ball, B.C., Jr.: Letter to the Author (Dec. 14, 1982).
2. Abouzeid, K.M.: "Corporate Goal Identification and Achievement," *Managerial Planning* (Nov.-Dec. 1980) 29, no. 4, 17-22.
3. Newman, W.H.: *Constructive Control — Design and Use of Control Systems*, Prentice Hall Inc., Englewood Cliffs, NJ (1975).
4. Roush, C.H. and Ball, B.C., Jr.: "Controlling the Implementation of Strategy," *Managerial Planning* (Nov.-Dec. 1980) 29, no. 4, 3-12.
5. Kiechel, W., "Corporate Strategists Under Fire," *Fortune* (Dec. 17, 1982) 106, no. 13, 34-39.
6. Ball, B.C., Jr.: "How Do You Get 'Operators' to Think Long Term?" Unpublished article.
7. Batten, J.D.: *Beyond Management by Objectives*, AMACOM, New York (1966).
8. Sawhill, J.C.: "How to Rescue Strategic Planning in Energy," *Oil & Gas Journal* (Feb. 28, 1983) 81, no. 9, 90-92.

Appendix A:

The Petroleum Data System

The Petroleum Data System (PDS) is a family of computerized databases (see fig. 1) containing oil and gas industry information compiled from publicly available sources. The major file in the system is called TOTL, which consists of composite reservoir and field information on oil and gas fields in the United States. This database contains most of the information required for the strategic analyses of fields and basins described in this book. The data presented is gathered from state agencies, Oil Scouts, and other special reporting units. The completeness of the data for particular areas depends upon the completeness of the corresponding source reports for those areas. Although reserve data are frequently not present in the file, volumetric parameters and annual production by field starting with 1968 are generally available, so decline or volumetric analyses can be conducted to estimate ultimate reserves.

An additional source for the kind of information used in this book is the AAPG Exploratory Well File (CSDE). This file is supported by the American Association of Petroleum Geologists and contain information on all exploratory wells drilled between 1954 and 1981. Records for successful field exploratory wells carry ultimate yield codes which provide rough estimates of the reserves available in new field discoveries. These codes can be used to check the volumetric or the production decline estimates obtained from the TOTL file. The CSDE file can also be used to evaluate the exploration history of a given basin.

This appendix discussing the Petroleum Data System and the American Association of Petroleum Geologists' Exploratory Well File was prepared specifically for this text by Information Systems Programs of the University of Oklahoma's Energy Resources Institute. The Exploratory Well File of the American Association of Petroleum Geologists (AAPG) Committee on Statistics of Drilling, is available for public access through the University of Oklahoma.

In addition to the publicly available data contained in the Petroleum Data System, most company analysts have access to proprietary data in their own files. These will include reserve estimates and reservoir parameters that may not have been reported by the state or other general reporting agency used as a data source for PDS. Where PDS has missing data these proprietary files may be consulted to provide a more complete data presentation.

In addition to the major files described above there are several additional files in the Petroleum Data System which may be of use for specialized studies. Of special interest, for example, may be the offshore, secondary recovery, and brine files. General descriptions of each of the files contained in the Petroleum Data System are included below.

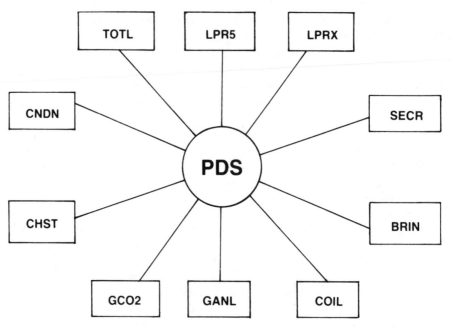

Figure 1.   Petroleum Data System.

U.S. Field/Reservoir Geological & Engineering Characteristics

The U.S. Oil and Gas File (TOTL)

This file consists of more than 100,000 records containing publicly available information for all fields and reservoirs in the United States. Data elements contained in the records may include identification of fields and reservoirs by name and code, location, and present producing status. Geological and engineering data may consist of the name and age of the producing formation, discovery method, trap type, drive, lithology, depth, acreage, spacing, thickness, porosity, permeability, gravity, pressure, and temperature. Also available is information on the discovery well of the field/reservoir.

The information contained in TOTL is obtained from many state publications, including state annual production reports, computerized tapes and key data from hearing files; Dwight's Energy Data; the International Oil Scouts Association; the U.S. Geological Survey; and the Energy Information Administration.

Each record in TOTL consists of general information on a specific field or reservoir. Each state is updated annually with production, well counts, new discoveries, and other available data. The state agencies vary in the way their data are reported. Each field or reservoir record is classified as being oil, gas, or associated gas, depending on how it is designated by the state. Some records have more than one classification. Offshore records are also maintained in these files. Oil and gas production is carried annually beginning with the year 1968. An all-time cumulative is also carried on fields or reservoirs when that information is available.

Secondary Recovery Information

The Secondary Recovery Projects File (SECR)

This database consists of information on secondary recovery projects. Presently, information is available on projects in the states of Texas, Kansas, Louisiana, and Illinois. Data may consist of location information, name of operator and lease, lithology, porosity, permeability, viscosity, acres, number of wells, pressure, and ultimate recoverable information. Also included are production data, information on the types of injection, and the fluids injected by the project.

Data on secondary recovery projects in Texas are gathered by a survey every two years conducted by the Texas Railroad Commission. The data on the other states are obtained as available from state publications.

The Oil Field Water Analyses File (BRIN)

This file contains information on more than 77,000 oil field water samples compiled from water analysis reports collected from oil companies. Each record consists of a single analysis taken at a specific well location. However, there may be more than one analysis available per well. The data are very useful for several types of studies including: enhanced oil recovery studies, geochemical exploration studies, petroleum exploration and production studies, and environmental studies.

Location information reported may include state, county, geologic basin, field, and section, township, and range. Well records may include information such as the company well name and well number, the API well number, the classification of the well, the deepest formation penetrated by the well, the age of the formation, the total depth of the well, elevation, subsea data, bottom hole temperature and pressure, and chemical constituents of the sample. Also, there is information reported on the area of the formation, age, depth interval, the porosity, permeability, pH, specific gravity, resistivity, lithology, and temperature and pressure of the interval at which the sample was measured. Also reported are the date the sample was taken and where it was taken.

The Natural Gas Analyses File (GANL)

This file contains routine analyses and related source data from natural gas samples. These samples were collected as part of a continuous survey of the free world for occurrences of helium. The database contains all samples released for publication through December 31, 1977.

Data elements reported may include the location, the name and the age of the formation, the dates sampled, the depth, thickness, well head pressure, specific gravity, and heating value, as well as the molecular percent of the gas constituents.

Each record consists of a single analysis taken at a specific location. Usually the sample is taken at a well, but there are also a few samples taken from pipelines and other locations. There may be more than one analysis listed per well.

The analyses have been collected by the Bureau of Mines since 1917. The data are now maintained by the U.S. Department of Energy.

The Carbon Dioxide Database (GC02)

This database contains 630 analyses from 27 states, and contains an extract of those analyses from the Natural Gas Analyses File (GANL) having a molecular percent of carbon dioxide greater than 5 percent. The database was developed by Gulf Universities Research Consortium (GURC) with the U.S. Department of Energy in an attempt to locate natural occurrences of carbon dioxide.

Each record consists of survey information supplied by a company for the year 1976. Reserves reported include those dedicated to interstate contract, intrastate contract, and company use. Also reported is the commitment status of the reserves, changes and totals of proved gas reserves by type of gas occurrences, net completions, gas production, and successful gas well footage drilled for the report year.

The U.S. Department of Energy currently maintains these data.

The Crude Oil Analyses File (COIL)

This file contains 8600 crude oil well samples for the United States. A few are for wells outside the United States. The analyses may include information on location, name of field, formation, age, depth, lithology, sulphur content, nitrogen, pour point, carbon residue, and percent of crude. The analyses also include volume percents of crude oil, viscosities, cloud points, and specific gravities.

The crude oil analyses were originally collected by the Bureau of Mines. The data are now maintained and furnished to the Petroleum Data System by the Department of Energy.

Federal Offshore Lease Information

The information described in this section is obtained from the Minerals Management Service.

The Lease Production Revenue System-5 File (LPR5)

This database contains lease bidding information including the names of bidders and amounts of all bids received on federal outer-continental shelf leases. The database consists of a single record for each bid on a lease. Lease information and location are duplicated on each record. Lease information common to each record includes the sale date, lease acres, track number, total blocks, block description, royalty rate, and type of lease. Bid information includes the number of bids, company name, acres, bid, and percent ownership.

The Lease Production Revenue System-10 File (LPR10)

This database contains annual production, royalty, and rental amounts for each federal outer-continental shelf lease. Each record consists of all years of rental, royalty, and production on a lease. The yearly information begins with 1952 and includes: lease number, state, area, acres, royalty rate, number of bids, number of rental years, number of production years, status of lease, lease-type, bid amounts, sale data, and lease data.

The data for this file are compiled and maintained by the Minerals Management Service.

Canadian Production and Reserve Files

The information contained in the CHST file is collected by the Energy Resources Conservation Board, Alberta, Canada. As for the CNDN file, the data for the Province of Alberta were supplied by the Geological Survey of Canada and the Energy Resources Conservation Board of the Province of Alberta.

The Canadian Historical Production/Injection File (CHST)

This file contains the amount of production and the amount of fluids injected for each year since 1962 for each oil or gas reservoir in the Province of Alberta.

These records consist of information on specific reservoirs or pools. Data elements include the types and amounts of production reported for the reservoir (e.g., NGL, propane, butane, pentane, crude oil, water) and the types and amounts of fluids injected into the reservoir (e.g., injection gas, steam, water). Total cumulative production for each type of product produced as well as for each type of injection is also available.

The Canadian Reserve File (CNDN)

This file contains information on oil and gas fields and reservoirs for the Province of Alberta. CNDN is unique in its detailed reserve information and complete reservoir data for all producing wells and pools in enhanced recovery areas. Western Canadian information is highly regarded for its completeness and quality. Historically, it has been more accessible to the public than comparable U.S. data.

Data elements may include the location, name of field and reservoir, formation, status, depths, recovery percentage, porosity, water saturation, temperature, and shrinkage. Data elements also may include original oil or gas in place, recoverable oil, and reserves.

Additional Associated Data File

The AAPG/CSD Exploratory Well File

The Exploratory Well File of the American Association of Petroleum Geologists (AAPG), Committee on Statistics of Drilling (CSD), is available for public access through the University of Oklahoma. The AAPG/CSD Exploratory Well File includes information on all exploratory wells drilled between 1954 and 1982. Basic well data pertaining to wells drilled in the United States are reported by means of an individual well ticket initiated by respondents to the American Petroleum Institute. The University of Oklahoma is the repository for the data.

Exploratory well classifications include new field wildcat, new pool wildcat, deeper pool test, shallower pool test, and outpost (extension) test. The types of information in a record are the American Petroleum Institute well number, completion date, depths, well classification, number of completions, operator identification, Section-Township-Range (where available), geological basin, deepest formation code, name, pay code and pay name, estimated ultimate yield, field name, and remarks. The database is indexed by state, year reported, basin code, and classification.

This database contains 313,470 well tickets for the years 1954–1982. Data for the years 1954–1963 were obtained by the University of Oklahoma as hard copy from the AAPG office in Tulsa, Oklahoma, and were data reduced and added to the exploratory well file under contract to the U.S. Department of the Interior.

How to Access

To access the Petroleum Data System and the additional data file, contact Information Systems Programs, Energy Resources Institute, University of Oklahoma, P.O. Box 3030, Norman, Oklahoma, 73070, 405/360-1600.

The Petroleum Data System
Oil and Gas Data from the TOTL File
Field Data from the Permian Basin in New Mexico

Figure 2 exemplifies the kind of data available in the TOTL data file. Data are organized by field and/or reservoir in TOTL. In order to generate this report, reservoir records are merged by field. The computer analysis takes reservoir data from the TOTL file and calculates "year discovered," "depth," "producing acres," "trap type," "discovery method," "cumulative production," "original in place," and "original barrels of oil equivalent" for the field. A field's "year discovered" is the median year of discovery for the reservoirs associated with the field; the field's "depth" is the median depth of the associated reservoirs. The field's "trap type" and "discovery method" reflect values that occur most frequently in the associated reservoir. The remaining data items for the field are summed from the associated reservoirs. The Permian basin is the only area for which TOTL contains data on "original oil in place," "original total gas in place," and "original barrels of oil

THE PETROLEUM DATA SYSTEM
OIL AND GAS DATA FROM TOTL FILE
FIELD DATA FROM THE PERMIAN BASIN IN NEW MEXICO

FLDCODE	FIELD	YRDISC	DEPTH	TRAP	DISCOVERY METHOD	CUMULATIVE OIL (BBLS)	CUMULATIVE TOTAL GAS (MCF)	ORIGINAL OIL IN PLACE (BBLS)	ORIGINAL TOTAL GAS IN PLACE (MCF)	ORIGINAL BOE IN PLACE (BBLS)
35-204920	DUFFIELD	1952	8616		SEISMIC	•	4421688	•	6190000	1105357
35-210184	E-K	1959	4944	LATERA		5998136	•	23914400	14666000	26533329
35-210215	E-K, EAST	1957	4838	LATERA	SUBSURF	1087817	•	6152710	3930000	6854496
35-210463	EAGLE CREEK	1975	4577			2250339	7191677	7459460	•	•
35-210494	EAGLE CREEK, EAST	1977	•			•	17545625	•	•	•
35-212447	EAVES	1929	3228	BIOHER	SUBSURF	6706253	•	•	•	•
35-212819	ECHOLS	1959	9446		SEISMIC	4562557	•	17570600	26005000	22014350
35-212850	ECHOLS, EAST	1957	12021	ANTICL	SEISMIC	548158	•	2092200	3078000	2641843
35-212881	ECHOLS, NORTH	1952	12057	ANTICL	SEISMIC	1217365	•	4646410	6851000	5869803
35-215958	EIDSON	1965	10705	LATERA	SEISMIC	2299613	•	8765190	12175000	10939297
35-215970	EIDSON, NORTH	1978	•			6	138517	•	•	•
35-216935	EL MAR	1959	4609			5453625	•	22546192	1597800	25399406
35-218279	ELBOW CANYON	1976	•			5799	•	•	•	•
35-002695	ELKIN, SOUTH	1981	•			59620	231237	•	•	•
35-219911	ELKINS	1981	•			529	•	•	•	•
35-220407	ELLIOTT	1952	7366	LATERA	OLD WEL	16548	•	876500	840000	1026500
35-223538	EMERALD	1967	9436			82618	82865	353648	317000	410255
35-224313	EMPIRE	1957	4545			20721044	29102268	391517761	183547000	424294011
35-224375	EMPIRE, EAST	1961	786			473582	•	1591700	174000	1622771
35-224382	EMPIRE, SOUTH	1974	•			432122	•	1892920	•	1892920
35-228258	ESPERANZA	1969	2042			454417	37015	1599600	740000	1731743
35-228661	ESTACADO	1963	11478		SEISMIC	11592	•	•	•	•
35-229808	EUMONT	1953	3457			•	1382012255	•	1606500000	286875000
35-229799	EUNICE, NORTH	1972	4113			•	890773	•	2990000	525000
35-229932	EUNICE, SOUTH	1969	3300			687048	915646	3196590	1274000	3424090
35-000131	EUNICE, SOUTHWEST	1980	•			4708	82135	•	•	•

Figure 2.   Example computer report of TOTL-field data.

equivalent in place." (These data result from a Department of Energy survey conducted several years ago on the Permian Basin.) For all other basins, a person doing strategic analysis must have access to additional information sources.

This type of analysis can be replicated for other areas, based on field rather than reservoir records. Field records on TOTL carry the "year discovered," which is the earliest year of discovery for reservoirs within the field; "depth" on a field record reflects the deepest reservoir in the field. Field records always carry annual and cumulative oil and gas production data.

The Petroleum Data System
Oil and Gas Data from the TOTL File
Field Production Data from the Permian Basin in Texas

Figure 3 shows five years of annual crude oil production data and additional information that would be required for doing a decline analysis to obtain ultimate yield estimates from data in the TOTL file. There are similar production figures for both associated and nonassociated gas. In this figure, reservoir oil production was summed to show a field total since production is carried on reservoir records in Texas.

For new field and reservoir discoveries a volumetric estimate of ultimate oil or gas may be more appropriate than decline analysis.

The Petroleum Data System
Oil and Gas Data from the TOTL File
Reservoir Data from the Permian Basin in New Mexico

Figure 4 is an example of reservoir data available from the TOTL file which is of special interest in strategic analysis. Most of the data in this figure are carried on reservoir (pool) records. Some states, including Oklahoma, do not report production by reservoir but only by field. In these cases, the cumulative oil and cumulative gas and annual oil and annual gas data are carried on field records only.

The Petroleum Data System
AAPG Exploratory Well Data
Well File Data from the Williston Basin

Figure 5 is a representative retrieval from the CSDE exploratory well file maintained by the American Association of Petroleum Geologists. This file is particularly

THE PETROLEUM DATA SYSTEM
OIL DATA FROM TOTL FILE
FIELD PRODUCTION DATA FROM THE PERMIAN BASIN IN TEXAS

FLOCODE	FIELD	YRDISC	TRAP TYPE	ANNUAL 1977 (BBLS)	ANNUAL 1978 (BBLS)	ANNUAL 1979 (BBLS)	ANNUAL 1980 (BBLS)	ANNUAL 1981 (BBLS)	CUMULATIVE OIL (BBLS)	ORIGINAL OIL IN PLACE (BBLS)
48-000753	A & K	1957	ANTICL						2797	*
48-000784	A & S	1968		7938	8311	7462	6884	5576	165322	804034
48-001063	A P	1958	ANTICL						1621	6957
48-001094	A P C	1959		11217	7735	6171	9876	8367	235882	*
48-000878	A. G. H.	1980					682	15551	19317	*
48-001218	A. W.	1964	ANTICL	18416	25505	24426	22658	20116	2147193	8950190
48-000832	A-B-M	1974							28918	110953
48-000879	A-F-G	1979				4061	15557	17686	37304	
48-001249	AAKER	1957							2570	10936
48-001652	ABELL	1961	UNCONF	753442	718419	670114	596776	561530	34416818	118554153
48-001683	ABELL, EAST *	1958	MONOCL	56820	47127	50222	57812	53211	4473248	22847686
48-001745	ABELL, NORTH	1961	ANTICL	3113	2904	2699	2497	2229	426625	1883527
48-001714	ABELL, NORTHEAST	1963	PINCHO	9024	8357	7909	7443	7578	354181	1495892
48-001776	ABELL, NORTHWEST	1955		7071	5131	3405	2949	3099	1493877	6846273
48-001838	ABELL, SOUTH	1951		24860	24731	22698	18854	15742	955434	3986968
48-001807	ABELL, SOUTHEAST	1962	FAULT	13785	10398	9039	6403	8913	948451	4229415
48-001869	ABELL, WEST	1964	ANTICL	598	589	531	499	403	46496	262113
48-002385	ACKER	1977		5160	4890	1903	8036	27376	47365	*
48-002427	ACKERLY	1954	LATERA	1483449	1510057	1658056	1490508	1445342	30521963	710765872
48-002458	ACKERLY, EAST	1955	BIOSTR						5995	21868
48-002489	ACKERLY, NORTH	1958	BIOSTR	66004	86057	80447	93138	86121	1582393	5543507
48-002520	ACKERLY, NORTHWEST	1960	BIOHER	17627	12699	9907	8969	6736	464974	2378240
48-002551	ACKERLY, SOUTH	1954							1080	7013
48-002799	ACORN	1965	REEF						10773	37932
48-002800	ACR	1978		264	17075	28133	16790	11978	7216	*
48-003109	ADAIR	1949	LATERA	4592428	4087025	3456889	2882057	2766639	93886925	453556016

Figure 3. Five years oil production from TOTL file.

THE PETROLEUM DATA SYSTEM
OIL AND GAS DATA FROM TOTL FILE
RESERVOIR DATA FROM THE PERMIAN BASIN IN NEW MEXICO

FLDCODE	FIELD	RESERVOIR	PRODUCING FORMATION	GEOLOGIC AGE	PRIMARY LITHOLOGY	YRDISC	DEPTH	TRAP	CUMULATIVE OIL (BBLS)	CUMULATIVE TOTAL GAS (MCF)
35-002675	ACME	SAN ANDRES	SAN ANDRES	PERM	DOLOMIT	1951	1975	NOSE	240819	•
35-005527	AID	MORROW	MORROW	PENN		1971	10608		•	2135550
35-005527	AID	YATES SEVEN RIV	YATES SEVEN RIV	PERM	DOLOMIT	1941	850	DOLO	153091	•
35-005883	AIRSTRIP	BONE SPRING, LO	BONE SPRING, LO		LIMESTO	1980	•		20440	9065
35-005883	AIRSTRIP	BONE SPRING, UP	BONE SPRING, UP		LIMESTO	1980	•		908940	413935
35-005883	AIRSTRIP	BONE SPRINGS	BONE SPRINGS	PERM	LIMESTO	1979	•		271955	100256
35-005883	AIRSTRIP	WOLFCAMP	WOLFCAMP	PERM	SANDSTO	1975	10656		119997	76430
35-006684	ALACRAN HILLS	ATOKA	ATOKA	PENN	SANDSTO	1978	•		•	464323
35-011851	ALLISON	ABO	ABO	PERM	DOLOMIT	1961	8970	ANTI	154413	•
35-011851	ALLISON	PENN	ROUGH	PENN	LIMESTO	1954	9670	ANTI	21625209	15315992
35-011851	ALLISON	SAN ANDRES	SAN ANDRES	PERM	DOLOMIT	1964	3000		118	•
35-011913	ALLISON NORTH	PENNSYLVANIAN	PENNSYLVANIAN	PENN	DOLOMIT	1957	9652		116038	•
35-011882	ALLISON, EAST	PENNSYLVANIAN	ROUGH C	PENN	DOLOMIT	1961	9659		20766	•
35-011882	ALLISON, EAST	SAN ANDRES	SAN ANDRES	PERM	LIMESTO	1977	•		8645	32648
35-011913	ALLISON, NORTH	SAN ANDRES	SAN ANDRES	PERM	DOLOMIT	1964	3000		848	•
35-011975	ALLISON, WEST	PENN	ROUGH C	PENN	LIMESTO	1962	9793		88667	•
35-017400	ANDERSON	PENN	BEND	PENN	SANDSTO	1954	11039		•	16014752
35-017400	ANDERSON	GRAYBURG	GRAYBURG	PERM	SANDSTO	1940	2520		1269659	•
35-017462	ANDERSON RANCH	DEVONIAN	DEVONIAN	DEVO	DOLOMIT	1953	13374	ANTI	8294455	2864248
35-017462	ANDERSON RANCH	MORROW	MORROW	PENN		1968	12160		•	8441117
35-017462	ANDERSON RANCH	PENN	PENN	PENN	DOLOMIT	1954	12362		167	•
35-017462	ANDERSON RANCH	WOLFCAMP	WOLFCAMP	PERM	LIMESTO	1953	9664	ANTI	3367224	•
35-017493	ANDERSON RANCH, EAST	PENNSYLVANIAN	PENNSYLVANIAN	PENN	LIMESTO	1957	10960	ANTI	10314	•
35-017493	ANDERSON RANCH, EAST	WOLFCAMP	WOLFCAMP	PERM	SHALE -	1980	•		36897	32296
35-017524	ANDERSON RANCH, NORT	WOLFCAMP	WOLFCAMP	PERM	LIMESTO	1962	8001		5635991	•
35-017555	ANDERSON RANCH, SOUT	WOLFCAMP	WOLFCAMP	PERM	SHALE -	1962	9660		51299	•

Figure 4. Example of reservoir data from TOTL file.

THE PETROLEUM DATA SYSTEM
AAPG EXPLORATORY WELL DATA
WELL FILE DATA FROM THE WILLISTON BASIN

10:47 TUESDAY, MAY 24, 1983

APIWELL	FIELD	OPERATOR NAME	OPRCODE	COUNTY CODE	SEC	TWNSHP	RANGE	ELEV	TOTAL DEPTH	DEEP FORM	FINAL LAHEE	WELL CLASS	COMP DATE	COMPLETE OIL	COMPLETE GAS	ULTIMATE OIL	ULTIMATE GAS
	BREDETTE, NORTH	CALIFORNIA CO.	795	019	34	033N	049E	•	7477		1	E	57/05	1		E	
2118500	BRORSON	BRINKERHOFF DRLG	140	083	8	023N	058E	•	12550	RDRV	5	E	77/09	1	0	E	
2121400	BRORSON	PETROLEUM INC	633	083	2	023N	058E	•	9700	MSNC	5	E	78/05	1	0	E	
2154200	BRORSON S	TENNECO OIL	840	083	11	024N	057E	•	12700	RDRV	5	E	80/05	0	0		
2147700	BRORSON S	KERRY PET	999	083	14	023N	057E	•	12581	RDRV	5	E	81/06	0	0		
21044	BRUSH MOUNTAIN	SINCLAIR OIL & GAS	755	091	3	032N	059E	2132	•	RDRV	5	E	68/07	0	0		
21049	BRUSH MOUNTAIN	SINCLAIR OIL & GAS	755	091	0	031N	059E	2206	•	RDRV	5	E	68/11	1	0		
2137400	CANAL	AMOCO PROD	615	083	5	023N	059E	•	12800	RDRV	5	E	80/11	0	0		
21017	CAT CREEK	CARNELL E F	999	069	1	015N	029E	2830	3665	CRLS	3	E	69/11	0	0		
2113700	CHARLIE CREEK	FARMERS UNION EX	302	083	24	025N	054E	•	11825	RDRV	1	E	76/12	1	0	E	
2115300	CHARLIE CREEK	FARMERS UNION EX	302	083	23	025N	054E	•	11680	RDRV	2	E	77/05	1	0	E	
2129600	CHARLIE CREEK	FARMERS UNION CE	302	083	23	025N	054E	•	11704	RDRV	3	E	79/05	0	0		
2149700	CHARLIE CREEK	MOSBACHER PRD	999	083	26	025N	054E	•	11720	RDRV	3	E	81/05	0	0		
2130900	CHINOOK	FULTON PRODUCING	999	005	21	033N	020E	•	1500	TPCK	3	E	72/05	0	1		E
2190400	CHIP CREEK	WISE OIL CO	999	005	22	026N	017E	•	1811	CRCS	5	E	77/02	0	1		E
2121100	CLEAR LAKE	PATRICK PET	619	091	17	033N	058E	•	8965	NSKU	4	E	78/12	1	0		
2145000	COMERTOWN	DIAMOND SHMRK	731	091	14	036N	057E	•	6850	RCLF	4	E	81/10	1	0		
2117400	DAGMAR	PATRICK PET ETAL	619	091	28	033N	057E	•	11430	RDRV	1	E	76/10	1	0	E	
2139700	DAGMAR N	BEREN CORP	999	091	11	033N	057E	•	11400	RDRV	5	E	81/05	0	0		
2124500	DIVIDE	ORRIT O&G CO	999	091	27	034N	058E	•	11500	RDRV	1	E	79/07	1	0	E	
21091	DWYER	MONSANTO CO	498	091	16	032N	059E	•	7981	RCLF	5	E	70/01	1	0		
2135000	EAGLE	BASS ENTRPRIS	097	083	28	022N	059E	•	10782	DPRW	3	E	80/05	1	0	E	
05089	FAIRVIEW	CONSOLIDATED OIL & G	999	083	2	025N	058E	2112	•	RDRV	5	E	66/12	0	0		
21012	FAIRVIEW	CONSOLIDATED OIL AND	999	083	3	025N	059E	2208	•	RDRV	5	E	68/04	1	0		
21060	FAIRVIEW	TENNECO OIL	840	083	29	025N	059E	•	12875	RDRV	5	E	70/10	1	0		
2140200	FAIRVIEW	AL-AQUITAINE	999	083	15	025N	058E	•	13300	MSNC	5	E	81/03	1	0	E	

Figure 5. Data from CSDE exploratory well file.

useful for evaluating exploration in a given area and has a large amount of important information in standard AAPG codes. Operator codes are contained in the AAPG handbook. The other codes are included on the page following the printout to make it easier to understand what is available in this file. The ultimate oil and ultimate gas are generally available on new field discovery records and often not on other records. These codes give general yield information and are not designed for detailed analyses. They do provide a valuable check against yield data obtained from other techniques.

Appendix B (part I):

Exploration Analysis Using Petroleum Information's Computer Data Bases

Petroleum exploration is perceived to operate as a three-part model that flows from basin analysis through play analysis to prospect definition. Except for frontier provinces that have few or no wells, computer data bases can be used as the primary analytical tool in this exploration model. An outline of procedures to evaluate basins, plays, and prospects using Petroleum Information's computer data bases is outlined on figure 1. A flow chart of the model is shown on figure 2. The analyses described in this model also support exploration planning and can be used to evaluate deals and submittals.

Analysts must establish objectives and guidelines for the evaluation. The selection of reports, maps, and statistics should be made in terms of the objectives. Success of the evaluation is measured in terms of how well the objectives are satisfied. Several objectives that might be considered are noted on figure 1.

The objective of basin analysis in this model is to identify areas or trends that offer the best exploration potential in respect to the established objectives. Drilling statistics derived from Petroleum Information's Drilling Activity Analysis System (DAAS) can be used to select favorable areas on the basis of high success ratios, high initial oil and gas potential and low drilling and completion costs per BOE. County summaries of annual exploration drilling statistics, shown on figure 3, provide the source for histograms and maps to facilitate interpretation. This report

This appendix (parts I and II) discussing P.I. proprietary digital data bases was prepared specifically for this text by Petroleum Information Corporation.

Procedures to evaluate basins, plays and deals using PI's
computer data bases:

Establish Exploration Objectives

1. Oil vs. Gas
2. Reserves
3. Strategies:
   a. Competitive Bid
   b. Onshore vs. Offshore
   c. Large Acreage Blocks, etc.
4. Financial Considerations

Basin Evaluation

1. Drilling Statistics by Province and County (Drilling Activity
   Analysis System - DAAS)
   a. Reports
   b. Maps
   c. Percent Success
   d. Average Oil & Gas IP
   e. Cost/BOE IP
2. Reserves by Field - (National Production System - NPS)
3. Economic Risk Analysis - Cost of Finding

Play Evaluation

1. Drilling Statistics by Formation - Reports and Maps (Well
   History Control System - WHCS)
2. Penetration and Show Maps - Define Degree of Exploration and
   Hydrocarbon Distribution
3. Reserves by Formation or Geologic Age (NPS)
4. Evaluate Lease Status (PI Computerized Ownership - PICO)
5. Economic Risk Analysis

Prospect Definition

1. Geologic Maps - Structure, Isopach, Residual (WHCS)
2. Deals and Submittals
3. Reserves by Reservoir (NPS)
4. Economic Risk Analysis

Figure 1.   Procedures to evaluate basins, plays, and deals using PI's computer data bases.

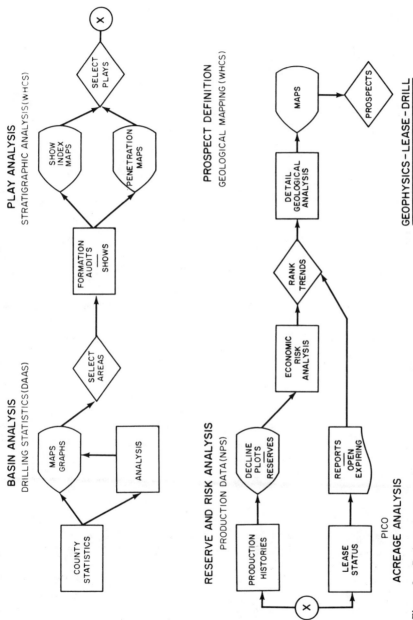

Figure 2. Basin evaluation flowchart.

County Analysis of Drilling Activity By Year
New Field Wildcat Wells - Permian Basin - 1974-1980

Chaves

YEAR	WELLS	WELLS	WELLS	WELLS	SUCCESS RATIO	DRILLED FOOTAGE	DRILLING COST	AVG FTG
1974		2	20	22	.09	84,713	2,308,889	3,851
1975	4	2	14	20	.30	122,669	4,053,540	6,133
1976	1	2	12	15	.20	67,603	2,270,258	4,507
1977	6	7	16	29				
1978	2	5	22	29	.24	182,118	9,207,171	6,280
1979	7	14	25	46	.46	212,895	11,813,375	4,628
1980	8	16	21	45	.53	214,775	13,961,435	4,773
County Total	28	48	130	206	.37	972,424	47,430,051	4,721

Chaves

YEAR	OIL IP BOPD	GAS IP MCFD	COND BCPD	OIL WELL AVG IP	GAS WELL AVG IP	TOTAL BOE/DAY	DRILLED FTG /BOE	DRILLED CT /BOE
1974		3,290	1		1,645	605	140	3,816
1975	150	1,834	1	38	917	488	251	8,306
1976	1	5,973	1	1	2,987	1,100	61	2,064
1977	119	20,089		20	2,870	3,813	23	1,001
1978	14	4,107		7	821	769	237	11,973
1979	225	11,383		32	813	2,317	93	5,099
1980	399	24,453		50	1,538	5,127	42	2,723
County Total	908	71,129	3	32	1,482	14,219	68	3,336

Figure 3. Drilling statistics in exploration — DAAS.

also can be generated by depth range for formation or geologic age to support play analysis. Summary by operators can be used to identify potential operating partners or candidates for acquisition.

Post oil embargo (1973 to date) exploration performance provides a good basis to forecast continued near term results. Qualified rankings—good, fair, or poor—of areas or trends can be made from the data in figure 3. Costs of finding and risk analysis are used to confirm the identity and ranking of the most favorable basins or areas. Drilling success and costs from this report are combined with resulting field reserves for determination of finding costs and economic risk analysis. With adequate production histories, well, reservoir, or field reserves can be extrapolated from decline plots (figure 4) from Petroleum Information's National Production System (NPS). Risk analysis at this stage should project the probability of finding X barrels with an N well program.

Stratigraphic plays and trends are evaluated from detailed drilling statistics such as the stratigraphic audit in figure 5, and penetration and show maps (figure 6) from PI's Well History Control System (WHCS). The maps help to define prospective trends by measuring the degree of exploration and hydrocarbon distribution. Reserves should be calculated by formation or geologic age for discoveries made during the period of the study (1973 to date is suggested). Detail risk analysis with generation of profit and loss projections is used to rank plays for exploration investment. Ranking of plays also is supported by analysis of acreage status and expiring acreage and ownership can be determined from reports as shown in figure 7.

In the model one would concentrate efforts to define prospects in the highest ranking plays or trends. Geological maps—structure, isopach, and residual—from WHCS (figure 8) would help to define prospects, leasing, geophysics, and drilling programs.

CRUDE AND GAS PRODUCTION DECLINE PLOTS

- Hard copy 8 1/2" x 11"
- 42X microfiche
- Date of first production
- Up to 12 years quarterly production
- Year end running cumulative
- Current year monthly production
- Rate VS time plots Crude Only
- Crude and casinghead gas/water where available

- Production by lease or well
- Number of wells on lease Gas Only
- Condensate/gas/water where available
- Production by well
- Upper and lower perforations
- Kansas
- North Arkansas (gas plots only)
- Louisiana
- Oklahoma

- New Mexico
- Texas
- Wyoming
- Southern Rockies (Arizona, Colorado, Nebraska, Nevada, Utah)
- Northern Rockies (Montana, North Dakota, South Dakota, Oregon)
- Semi-annually

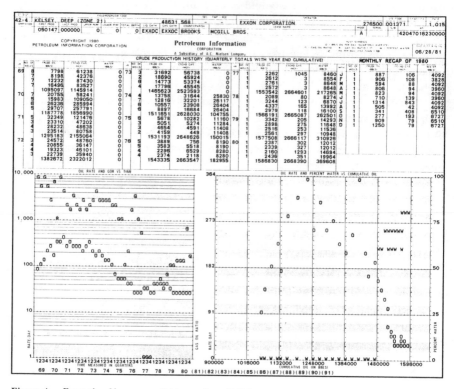

Figure 4.   Example of lease or well history from N.P.S.

Wind River Basin
Exploratory Well Data
State County - Natrona, Wyo

SEQUENCE	TOTAL PENETRATIONS	OIL	GAS	DRY PENETRA	SHOWS	PERCENT SUCCESS	SHOW INDEX	AVERAGE IP OIL BOPD	AVERAGE IP GAS MCFD	AVERAGE IP COND BCPD
Unknown	71	0	1	70	19	1.40	.27	0	1	0
Tertiary	726	0	0	726	1	.00	.00	0	0	0
Wind River	726	0	0	726	1	.00	.00	0	0	0
Wasatch	725	0	0	725	0	.00	.00	0	0	0
Ft. Union	725	2	1	722	13	.41	.01	24	8284	0
U Cretaceous-Lance	718	1	3	714	8	.55	.01	200	762	0
Lewis	694	0	0	694	0	.00	.00	0	0	0
Mesaverde (TPOT-PRKM)	691	1	0	690	9	.14	.01	670	0	0
Cody-Steel	685	1	1	683	23	.29	.03	30	2000	0
L U Cretaceous-Niobrara	643	0	0	643	5	.00	.00	0	0	0
Frontier	618	7	4	607	41	1.77	.06	134	1307	0
L Cretaceous-Mowry	575	0	0	575	4	.00	.00	0	0	0
Muddy	571	4	1	566	44	.87	.07	284	4364	0
Skull Creek	528	0	0	428	4	.00	.00	0	0	0
Dakota	500	5	0	495	36	1.00	.07	128	0	0
Lakota	446	3	0	443	22	.67	.04	88	0	0
Jurassic-Morrison	339	0	0	389	9	.00	.02	0	0	0
Sundance	273	0	0	273	2	.00	.00	0	0	0
Triassic-Nugget	259	1	0	258	6	.38	.02	415	0	0
Permian-Embar	224	0	0	224	4	.00	.01	0	0	0
Phosphoria	214	2	0	212	16	.93	.07	255	0	0
Pennsylvanian-Tensleep	211	6	0	205	43	2.84	.20	139	0	0
Amsden	30	0	0	30	0	.00	.00	0	0	0
Mississippian	23	0	0	23	0	.00	.00	0	0	0
Pre Cambrian	7	0	0	7	0	.00	.00	0	0	0
Totals/Average		33	11		310			172	2576	0

Wells Processed   830

Figure 5. WHCS stratigraphic audit.

133

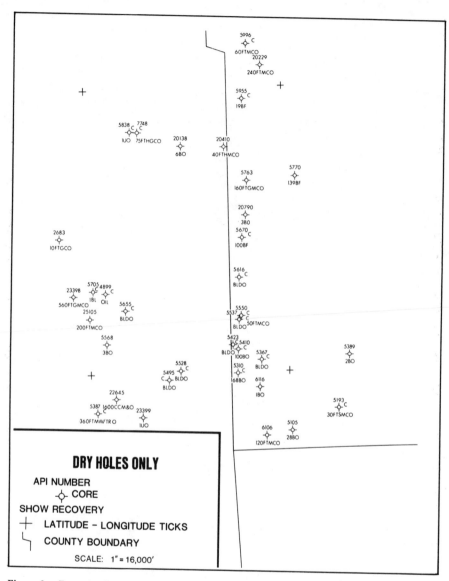

Figure 6. Example oil show map WHCS file data.

PICO RETRIEVAL - STATE: WY   DATE: 05/20/82

TOWNSHIP ACREAGE ANALYSIS
BY LEASE TYPE AND OPERATOR

MAP NO.	ST/CNTY	PROVINCE	TWNSP.	RNG.	SEC.	ACRES	IRRG-AC	LAND STATUS	LEASE NAME/NO.	HOLD CODE	EXP. DATE
W114	49 025	276	36 N	86 W	29	0040		P	Operator G	Unt.	
W114	49 025	276	36 N	86 W	29	0240		F	0	Unt.	
W114	49 025	276	36 N	86 W	29	0120		P	0	Unt.	
W114	49 025	276	36 N	86 W	29	0040		F	Operator L 091205	Unt.	
W114	49 025	276	36 N	86 W	30	0280		P	Operator X 012705	Unt.	
W114	49 025	276	36 N	86 W	30	0040		P	Operator X 012705	Unt.	
W114	49 025	276	36 N	86 W	30	0080		F	Operator G	Unt.	
W114	49 025	276	36 N	86 W	30	0040		F	Operator G	Unt.	
W114	49 025	276	36 N	86 W	30	0040		F	0		
W114	49 025	276	36 N	86 W	31	0160		F	Operator V 011230		04/83
W114	49 025	276	36 N	86 W	31	0040		F	Operator W 052565		04/83
W114	49 025	276	36 N	86 W	31	0160		P	Operator Y 104181		02/84
W114	49 025	276	36 N	86 W	31	0360		P	0		
W114	49 025	276	36 N	86 W	31	0040		F	0		
W114	49 025	276	36 N	86 W	32	0080		F	Operator Z		09/89
W114	49 025	276	36 N	86 W	32	0080		F	0		
W114	49 025	276	36 N	86 W	32	0280		P	0		
W114	49 025	276	36 N	86 W	32	0200		F	Operator A 078561		02/91
W114	49 025	276	36 N	86 W	33	0640		F	Operator A 078561		02/91
W114	49 025	276	36 N	86 W	34	0080		P	Operator T 007708		01/87
W114	49 025	276	36 N	86 W	34	0160		P	Operator T 007708		01/87
W114	49 025	276	36 N	86 W	34	0200		P	Operator AA045324		02/80
W114	49 025	276	36 N	86 W	34	0120		P	Operator AB108335		02/80
W114	49 025	276	36 N	86 W	34	0080		P	Operator C 091205		01/85
W114	49 025	276	36 N	86 W	35	0160		F	Operator T 007708		01/87
W114	49 025	276	36 N	86 W	35	0320		F	Operator F 014702		07/83
W114	49 025	276	36 N	86 W	35	0160		P	Operator L 091205		01/85
W114	49 025	276	36 N	86 W	36	0480		F	Operator F 014702		12/86
W114	49 025	276	36 N	86 W	36	0160		F	Operator A 078561		02/91

N 36 W 86

TOWNSHIP/RANGE ACRES:	22,440	FEDERAL:	17,040	STATE:	880	PRIVATE:	4,520
OPEN:	2,400	EXPIRED:	320	INDIAN:		OPEN:	1,520
		OPEN:	880				

Figure 7. Computer generated lease operator report.

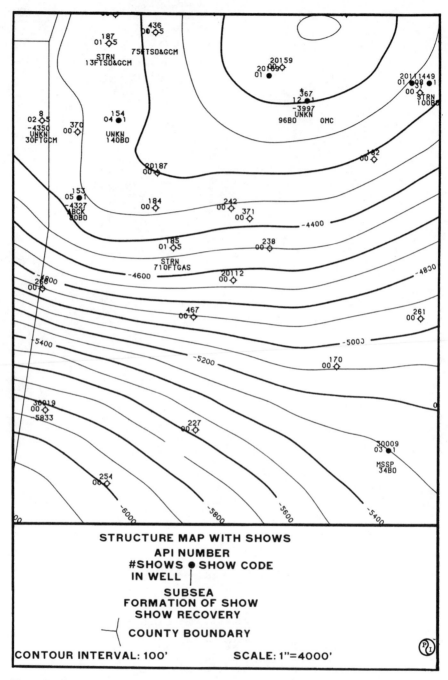

Figure 8. Example structure map, WHCS file.

Appendix B (part II):

Petroleum Information
Integrated On-Line Data Base System

The exploration analyses described in part I of Appendix B are supported by PI's integrated on-line data base system. The largest single array of U.S. petroleum exploration and production data bases are made available through this system. Except for the communication network all components of the system are supplied and maintained by Petroleum Information. Contracts for access to the system may be obtained from PI on-line and sales representatives. Each user is provided with a unique user number, documentation, and introductory training. Local telephone access to the system is provided through the Telenet communication system. Transmission rates of 300 and 1200 Baud are supported. Data are stored on disc and are available for real time direct access. CCA's Model 204 data base management system controls data base integration, the generation of packaged reports, and provides ad-hoc retrieval capability. Queries are made through easy to use conversational driver programs. Multiple files can be addressed during a single session. Users can generate reports and printer plots and control mapping, graphics, and analytical programs including economics risk analysis, from simple teletype terminals. Work files also can be transmitted for proprietary application processing on graphics work stations. PI's Computer Research graphics work stations include communication hardware and software interface to the on-line system. Monthly billing covers charges for data, processing, software, and transmission for each project. Information on prices, current status, and capabilities can be obtained from PI sales representatives or from On-line Services, Petroleum Information Corporation, P.O. Box 2612, Denver, CO 80201, (303) 740-7100.

PI's on-line data bases include Fast Permit Data, Active Well Data, Well History Control System (WHCS), Drilling Activity Analysis System (DAAS), National Production Data (NPS), and Computerized Ownership Data (PICO). Additional files to be added include rig availability, a Field Reserve file, and digital base maps. Data content, file coverage, and an outline of available reports are summarized in figure 1. Sample reports from some of these files are included in part I.

Fast Permit Data

The Fast Permit data base provides on-line access to current information on drilling permits which have been approved and released by regulatory agencies. These data are available within 24 to 48 hours from the time the approved permits are released to the public.

The Permit Data file contains the following elements

State
County
Location Data
Well Name
Well Number
Projected Depth
Projected Formation
Initial Class
First Report Date

Active Well Data

Active Well Data provides on-line access to current well data. Well data are entered upon release of permit and activity is tracked through completion. Daily drilling information on exploratory wells is accumulated and added directly to the data file. The information remains accessible through Active Well Data for 13 months following last reported activity. Active well data for the West Coast, Rocky Mountains, Northeast U.S., and Mid-continent are available in 1983. Gulf Coast and Permian Basin data will be added in 1984.

The Active Well Data file contains the following elements:

General Information
API Number
Well Location and Elevation
Operator Name
Lease Name
Field Name
Dates (start, suspension, etc.)

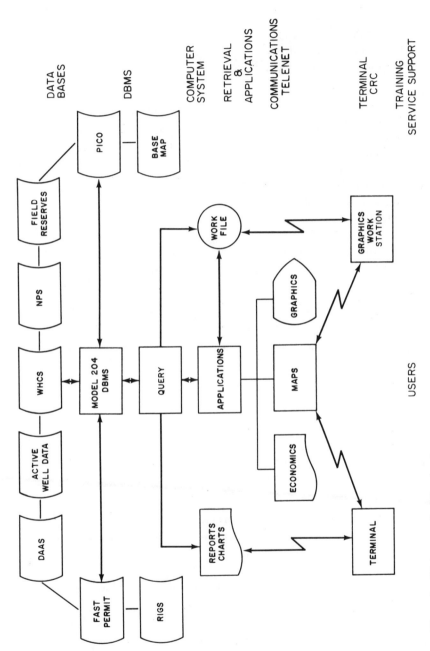

Figure 1. PI integrated on-line data base system.

Activity Data
Initial Potential Tests
Formation Tops and Bases
Drillstem and Wireline Tests
Production Tests
Core Data

Miscellaneous Data
Logs Run
Drilling Shows
Drilling Media
Casing Report
Bit Records
Drilling Narrative
Hole Deviation
Bottom Hole Temperature
Treatments
Volumes
Total Depth
Final Status

Report modules are summarized below. Active Well Data also can be merged with Historic Well Data for mapping.

Active Well Data Reports

Report Number	Report Name	Description
1	One-Liner	Quick reference for well identification
2	Header Data	Summary of well information—scout ticket data
3	Full Well Report	All information available
8	Formation Top List	Provides reported formation tops and their depths. Also includes API number, section, township, and range
10	Location List	Quick reference to well location
18	Casing/Liner/Tubing	Provides casing, liner, and tubing information
19	Active Drilling Narrative	Provides operator name, address, PI map I.D. numbers, drilling progress data, and completion data

20	Core Analysis	Provides just core analysis
21	Initial Potential Tests	Provides just initial potential tests
22	Production Tests	Provides just production tests
23	Cores	Provides just core narrative
26	Formation Tests	Provides general information on formation tests
29	Miscellaneous Data	This report provides miscellaneous data not included in the other reports
C	IC-Data Record	This report provides a cross reference of IC number with API number

Historical Well Data

Historical Well Data provides on-line access to well completion records on more than 1.5 million wells. The information is continuously edited and updated monthly to ensure a complete and accurate data file. Well information may be selected by a wide range of user selected criteria and retrieved in the form of reports and graphic displays including posted and contoured maps.

The Historical Well Data file may contain the following elements:

General Well Information
API Number
Well Location and Elevation
Latitude and Longitude
Operator Name
Lease Name
Field Name
Dates (start, suspension, etc.)
Core Data
Formation Tops and Bases

Test Data
Initial Potential Tests
Drillstem and Wireline Tests
Production Tests

Miscellaneous Data
Logs Run
Drilling Shows
Drilling Media
Bit Records

Drilling Narrative
Hole Deviation
Bottom Hole Temperature
Treatments
Volumes
Total Depth
Final Status
Casing Report

Report modules are summarized below.

WHCS Reports

Report Number	Report Name	Description
1	One-Liner	Quick reference for well identification
2	Header Data	Summary of well information — completion card data
3	Full Well Report	All information available
4	Lat/Long List	Provides latitudes and longitudes
5	Quikmap	Graphic displays of well locations
6	Lat/Long Window	Useful in mapping
7	Isopach List	Provides thickness of formations
8	Formation Top List	Provides depth and show codes of formations
9	Log Survey List	Lists logs run on well
10	Location List	Quick reference to well location
11	Formation Show List	Formation types, penetration, and show codes
20	Drillstem Tests	Provides just drillstem tests
21	Initial Potential Tests	Provides just initial potential tests
22	Production Tests	Provides just production tests
23	Cores	Provides just core data
24	Driller Shows	Provides just driller show information

National Production Data (NPS)

The Production Data file contains current and historic production information on about 96 percent of U.S. oil and gas production. (Northeast U.S. data are not covered.) The data are available as monthly production volumes as well as cumulative production totals from inception. Production test data and field/reservoir summary data are also available.

General Well Information

API Number
Well Location (as required by regulatory agencies)
Operator Name
Lease Name
Field Name

Production Information

Operator Production
Field and Reservoir Historical Production
Production Dates
Lease Historical Production
Depth of Production
Reservoir Summary Data
Field Summary Data
Water Production
Field/Reservoir Gravity

Miscellaneous Data

Test Data
Geologic ID of Reservoir
Geologic Formation Codes
Well Count Data

Report modules are summarized below. Graphics output of production decline plots on leases and wells and P/Z plots for gas may be provided. Maps of well and lease production and well test data also can be generated in combination with WHCS.

NPS Reports

Report Number	Report Name	Description
2	Operator Activity District	This report provides data on every reservoir produced by an operator.
3	Operator Field Total	This report provides production data on *all* fields within a district where an operator produced.
4	Operator Field Reservoir	This report gives a reservoir by reservoir production account of an operator in a particular field.
5	Operator Reservoir Lease	This report provides production data on each lease in a selected reservoir by operator.
7	Lease Well List	This report lists all of the API numbers of wells in a selected lease.
8	Lease Test Data	This report allows the user to retrieve *all* the tests or a given number of tests from each well in the lease.
9	Well Test	This report lists all tests for a given well within a lease.
10	Ledger	This report provides the Historical Production of an entity, including changes in lease operators.
11	Short Ledger	This report will generate the three most recent years of Historical Production of an entity, including changes in lease operators.
12	Ledger Time Window	This report will generate the Historical Production of an entity, including changes in lease operators within a user selected time frame.
13	Multiple Ledger	This report will generate the Historical Production for up to fifteen entities, within a user selected time frame.

Drilling Activity Analysis System — DAAS

The Drilling Activity Analysis System (DAAS) provides access to U.S. drilling statistics for well completions 1970 to date. The system contains detailed data concerning well identification, drilling costs, initial potential, final status, and success ratios. Well data from permit through completion are included. Data files are updated weekly. More than thirty report modules are provided.

Data elements available in the various reports are:

API Number	Gas and/or Oil
Class, Final	% of Development
Class, Initial	% of NF Wildcat
Cost per Foot	% of Other Expl.
Cost IP BOE/Day	% of Total
County API Number	Average Depth
County Name	Average IP
Depth	Cost
Depth Range	Footage
DEV, NFW and/or OEX % of Footage	IP
Cost	IC No. (PI Internal)
Count	IP Average Gas
Footage	IP Average Oil
Dry % of Development	IP BPOD
Dry % of NF Wildcat	IP Condensate
Dry % of Other Expl.	IP Equivalent Gas
Dry % of Total	IP MCFPD
Dry Average Depth	IP Oil Well Gas
Dry Cost	IP Equivalent/Day
Dry Footage	IP Equivalent/Foot
Dry Well County	Loc Data, Northeast
Field Name	Loc Data, Offshore
Field Number	Loc Data, T/R/S
Footage	Loc Data, Texas
Footage per BOE	Operator of Record
Formation at Prod.	Province AAPG Number
Formation at TD	Province Name
State API Number	Total Footage
State Name	Total Gas Well Count
State, Sub Name	Total Liquids
Status, Final	Total NFW Well Count

Success Ratio by Depth
Success Ratio by Development
Success Ratio by NF Wildcat
Success Ratio by Other Expl.
Success Ratio by Total
Total Average Depth
Total by Depth Range
Total Cost
Total Develop. Well Count
Total Dry Well County
Total Equivalents

Total OEX Well Count
Total Oil Well Count
Total Well Count
Value Gas
Value Oil
Value Oil & Condensate
Value Total
Well Name
Well Number
Equivalent BOE/Day

Appendix C:

Social Risk Management*

The Social Environment as a Risk Factor

The social environment is a factor which is at times overlooked in the evaluation of potential exploration areas or development of discoveries. Due to the dispersed nature of oil and gas exploration and the diversity of company involvement in any given area, the oil industry has largely been spared from the requirements of formal environmental impact statement (EIS) procedures (the most notable exceptions are major federal leasing programs) and hence from evaluating the compatibility of their exploration and development plans with any given social environment. Widespread trends in increased citizen awareness and involvement, however, are calling for renewed attention to the social environment in all phases of oil and gas development.

It is increasingly common for resource development projects to be delayed or canceled when citizens' issues are not identified and addressed early in the project planning process. Although this is an increasing trend, it is not a new phenomenon. Figure 1 displays the capital lost on large-scale projects because of delay or cancelation due to social, environmental, or political factors. This study was performed by Petro-Canada in 1978 and shows that some projects have lost up to 100 percent of the total project capital costs [1].

Oil industry managers who do not understand or appreciate this phenomenon are placing themselves at the risk of finding their projects "held hostage" to

---

*As developed and practiced by FUND Inc., Denver, Colorado. This appendix was prepared by FUND Inc. of Denver, Colorado.

Project	Lost Capital	Total Project Capital Costs	Percent of Lost Capital to Total Project Capital Costs
Trans-Alaska Pipeline	$800 million	$10.6 billion	8 percent
Garrison Diversion Unit	$400 million	$632 million	63 percent
Staten Island LNG Terminal	$35 million	$35 million	100 percent
Pt. Conception LNG Terminal	$150-180 million	$932 million	16-17 percent
Dodds/Roundhill Coal Mine	$5 million	$200 million	2.5 percent
Birch Island Uranium Mine	$1.5-3 million	$20 million	8-15 percent
Bathhurst Peninsula Exploration Program	$15 million	$15 million	100 percent

Figure 1.   Capital lost on projects cancelled or delayed due to social, environmental, or political reasons.

mitigation demands. These demands can amount to hundreds of thousands or even millions of dollars — dollars which were not recognized as possible project costs at the outset. Even in situations where formal EISs are conducted under the auspices of some responsible government entities such as the Bureau of Land Management or Bureau of Indian Affairs, managers can no longer afford to shut their eyes, even temporarily, to the external social environment and the implications it may have for the successful completion of their project. All too common are examples of citizen reaction, lawsuits, and even rescinded agreements, after millions of dollars have been expended in obtaining valuable — but undevelopable — lease rights.

Social Risk Management: The Process

Social risk management is a system designed to assist industry, business, and government decision makers in managing this new external environment as a way to protect their hard dollar investments. It was developed to minimize risk, contain costs, meet schedules, improve profits, and ensure long-term stability and productivity. This process demands a different range of project planning activities. It requires more up-front investment in time and personnel to evaluate the social risk and modify project plans. End run social costs and project disruption, however, can be considerably reduced.

Ideally, social risk management is an ongoing decision making process which should be applied well before initial exploration decisions are and continue throughout an exploration and development program. Figure 2 illustrates the steps in this ongoing process. It is designed to identify, evaluate, respond to, and monitor the public issues arising from exploration or production activities. It can, however, be applied at any of three levels or phases during project development: (1) social resource mapping is a technique used for preliminary social evaluation of potential development areas; (2) situational assessments are used to analyze specific sites for potential social risks and to identify the attendant costs required for successful project completion; (3) social impact management is used throughout the development phase to identify appropriate project modifications and/or mitigations and to ensure smooth project implementation.

Social Resource Mapping. The social resource mapping process is used to match gross level social data with likely areas of exploration interest. Social and cultural factors such as settlement patterns, work routines, and history of resource development trends are used to evaluate the compatibility of broad geographic areas with project development.

Two social resource maps are shown in figures 3 and 4. The larger geographic units in figure 3 are Social Resource Units (SRUs). This set of units was developed by FUND Inc., through work with the U.S. Forest Service, to assist them in understanding the effects of their management activities on community areas [2]. In this process the Forest Service was the first government agency to recognize that the continued use of traditional political and administrative boundary systems were leading to conflicts and costly errors in resource management [3].

In contrast, social resource maps delineate the natural human boundaries rather than artificial political designations of state, county, or other district jurisdictions. Within these human geographic units people share distinct histories, cultures, lifestyles, or economic pursuits. While none of these units are entirely homogeneous with regard to any of these traits, the blend within a given unit is unique and gives an area its distinctive social character. It is this character that shapes the concerns people have about their environment and the issues that will arise concerning a development project.

The SRUs result from an aggregation of data from Human Resource Units (HRUs) as shown in figure 4. HRUs represent a finer level of analysis for projects that will have a smaller sphere of influence or impact. Projects in urban areas often need to start even one step lower at "neighborhood" units. An analysis of the

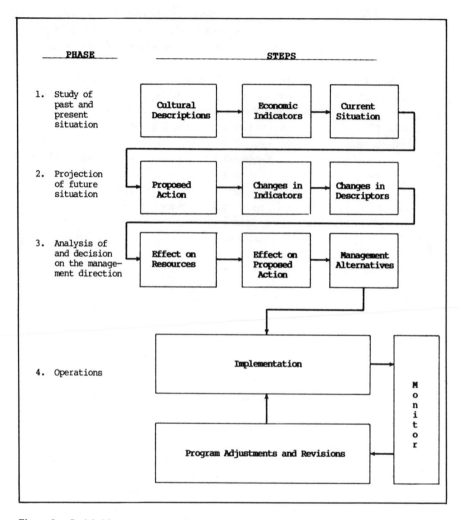

Figure 2.   Social risk management process.

social data for units of interest to developers will yield a ranking from areas of high compatibility, low conflict, and therefore low risk, to areas of low compatibility, high conflict, and therefore high risk [4].

The descriptive data that was used in delineating the units is available for all of the units mapped. Variables described in this data base are shown in figure 5.

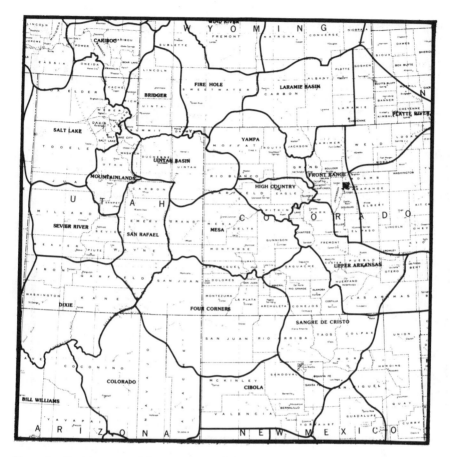

Figure 3. Social resource unit map.

FUND and associated organizations are currently in the process of improving and expanding this data base and related accessing systems, including the use of computer storage and retrieval. Other variables to be added to this data base include political structure and organization, historic economic activity, and additional demographic statistics. In addition, salient issues are monitored for current interpretations as well as future trend projections.

Situational Assessments. The situational assessment includes a community description process which documents local customs, routines, and informal community processes. This information is combined with more structured interview

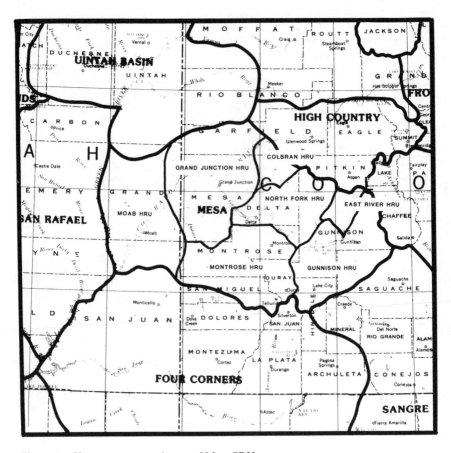

Figure 4.   Human resource unit map of Mesa SRU.

data. Both formal groups and informal networks are accessed in order to develop a well-balanced understanding of local community concerns and issues. This ensures that the horizontal decision making system of a community or group of communities is understood as well as the formal, more obvious decision making groups.

Understanding the nature of a community's horizontal networks is essential in evaluating the extent of potential risk to the project and how to reduce that risk. These horizontal networks are the informal groupings of people who often do not

Type of Variable	Variable Number	Variable Name
Cultural Descriptors	1	Existing/future publics
	2	Settlement patterns
	3	Work routines
	4	Communication networks
	5	Supporting services
	6	Recreational activities
	7	Geographic boundaries
Economic Indicators	8	Population change
	9	Employment mix
	10	Wage structure
	11	Local labor supply

Figure 5.   List of eleven key social variables for social resource units.

keep informed about pending projects, but who will react strongly if they feel their environment is threatened. Frequently, the formal (vertical) and informal (horizontal) social systems of a community are widely divergent. When this is the case, projects which have received formal sanction by conforming to all requirements and permitting procedures may still be in jeopardy.

Contacting both these formal and informal groups and understanding their concerns will ensure that the potential effects of the development are clearly evaluated and that there will be no surprises [5]. At this point, the exploration and production staff can consider what the potential *social costs* are of a development in this location. From an informed base, they can intelligently ask and answer questions such as the following: What will need to be included in the terms of the lease?

What kind of special conditions exist? Will traditional exploration, drilling, or transportation practices need to be modified? Where will potential conflict come from? Can that conflict be reduced or eliminated by actions the company can take? How much will that cost? In terms of these social risk factors, is the investment still justified or better placed in some other location?

In times of scarce capital, wise investments are even more critical. This kind of evaluation can help managers determine both where to place their leasing investments and where to allocate their development effort. This process can be used to develop new holdings or to evaluate current holdings for the "best" (i.e., most successful, least costly) development potential.

Social Impact Management. Social impact management is the ongoing process of communication and interaction between industry, citizens, and government groups throughout project development. As an exploration or development project is initiated, the knowledge of issues, concerns, formal groups, and networks derived from the situational assessment is used to guide the development of mitigation strategies. These strategies are developed with both government representatives and the citizen networks involved. When this process of early identification of issues and concerns occurs, fewer mitigations are required. In addition, many are not "extras" to which the developer must commit, but merely standard operating procedures which need to be explained and clarified to concerned groups.

Social impact management is actually a process of issue-centered planning and action, one which involves the informal citizen networks as well as formal groups in understanding the impacts of the proposed development and agreeing on the satisfactory resolution of concerns. This process is initiated as soon as the decision is made to proceed with development in a given area, and it should be continued by alert project managers throughout the project life to ensure continued project viability.

Summary

The social risk management system described here is a comprehensive approach to planning and action based on current and validated data. The approach minimizes local disruption and facilitates successful development. While a greater initial investment is required in this approach in order to carry out the required social mapping, perform related analyses, and then conduct situational assessments, several benefits are made available to the user.

The primary benefit, of course, is the prospect of reduced cost. Costs are reduced first of all by selecting sites with the highest degree of compatibility and therefore the lowest mitigation costs. Secondly, they are reduced by eliminating surprises which lead to project delay and/or more costly forms of confrontation such as litigation.

The third cost savings is in less community disruption. By working closely with government, interest groups, and citizens, project plans and mitigations are developed which are tailor made to the local area. Costly and unsuitable models need not be transplanted to "mitigate." Impacts can be managed internally by the community at much lower cost and with more successful outcomes if citizens are brought into the planning and development process at an early stage.

Today, social risk management is a necessary and critical part of the management process. Attention to the social environment is no longer a frill or a matter of social conscience. The oil and gas industry has traditionally assumed its projects would be welcomed. Without attention to local social issues and concerns, however, lease bonuses, rents, royalties, and promises of jobs will no longer be enough. Experience in the Overthrust Belt, in Alaska, off the Santa Barbara Coast, and on Indian reservations has shown that lack of attention to the social environment and to both formal and informal levels of communities will become an increasingly costly omission in project planning and development.

For additional information, write to or call:
The Foundation for Urban and Neighborhood Development (FUND), Inc.
2653 West Thirty-Second Avenue
Denver, Colorado 80211
(303) 433-7163

References

1. Loucks, Diane E., Perkowski, Dr. J., and Bowie, Douglas B.: *The Impact of Environmental Assessment on Energy Project Development*, Petro-Canada (1978).
2. Kent, James A., Greiwe, Richard J., Freeman, James F., and Ryan, John J.: *An Approach to Social Resource Management.* Prepared for USDA-Forest Service, Surface Environment and Mining (SEAM), Billings, MT (1979).
3. Findley, Rowe: "Our National Forests: Problems in Paradise," *National Geographic*, (Sept. 1982) 162, no. 3, 337–338.
4. Kent, James A. and Greiwe, Richard J.: "The Social Resource Unit: How Everyone Can Benefit from Physical Resource Development," *1978 Mining Yearbook*, Colorado Mining Association.
5. Mithen, B.J.: "Social Impact Analysis: Managing Surprise." Printed in *CIRC '81*, Annual Report of the Center for International Research Cooperation (CIRC), Commonwealth Scientific and Industrial Research Organization, Australia.

Index

Strategic Planning Associates, 16
Strategic planning phases, 9-14, 108; formulation phase, 9-11, 108; implementation phase, 12, 108; organization and control phases, 12-13, 108; reformulation — feedback and review phase, 13-14, 108
Strategies. *See* Corporate strategies; Development strategies; Exploration strategies; Financial strategies; Flexibility of strategies; Operating strategies
Strategy formulation: and planning scenarios, 103; for control system, 108; for strategic modeling, 83-84
Strategy implementation, 107-8
Success ratio. *See* Chance of test well success

Taylor, D.C., 36
10-K reports, 37-38
Tippee, Bob, xii
Top down planning, 19, 100, 106
Trial and error approach to exploration, 63
Trigger points (for contingency planning), 105, 109

Uniqueness in industry, vii
U.S. Geological Survey, 115
U.S. Oil and Gas File (TOTL), 113-15; for Permian basin, 119-23
Utility curves, 57

Vancil, R.S., 9
Venture participation, 53. *See also* Viability analysis
Viability analysis: and portfolio approach, 53, 56, 59; definition of, 42-43; examples of, 43-44, 50-52; multiple successes, binomial process, 44-46; multiple successes, true ruin case, 46-50; risk sharing in, 51-52; simulation model for, 48-50, 52-53; strategic considerations of, 53
Volumetric/geologic analog forecasting, 73; advantages of, 78; definition of, 74; limitations of, 78; main steps to, 74-75; prerequisites of oil accumulation for, 75-76

Water analysis, 116
Well History Control File (WHCS), 130, 133, 137, 141-42
Wiorkowski, J.J., 70
Working interest: as decision variable, 81; maximum allowable, 51-53; in viability analysis, 44
"WOTS-UP" analysis, 11

Zeger, S.L., 70
Zones within plays, 87-88